SpringerBriefs in Business

More information about this series at http://www.springer.com/series/8860

Siqing Shan · Qi Yan

Emergency Response Decision Support System

 Springer

Siqing Shan
School of Economics and Management
Beihang University
Haidian District, Beijing
China

Qi Yan
School of Economics and Management
Beihang University
Haidian District, Beijing
China

ISSN 2191-5482   ISSN 2191-5490   (electronic)
SpringerBriefs in Business
ISBN 978-981-10-3541-8   ISBN 978-981-10-3542-5   (eBook)
DOI 10.1007/978-981-10-3542-5

Library of Congress Control Number: 2016961273

Printed on acid-free paper

This Springer imprint is published by Springer Nature
The registered company is Springer Nature Singapore Pte Ltd.
The registered company address is: 152 Beach Road, #22-06/08 Gateway East, Singapore 189721, Singapore

Acknowledgements

We would like to express our gratitude to all those who helped us during the writing of this monograph.

First, we wish to thank all the members of our research team who participated in the early work, for their cooperation and hard work. Their names are Mengni Liu, Tong Zhang, Yinfeng Hou, Xiao Lin, Jihong Shi, Jie Ren, Zhilian Liu, Zhonghui Mao, and Tenglong Xin. The monograph is the crystallization of the wisdom of all the team members.

We wish to thank all the authors of the references cited in this monograph for their keen insight and rigorous attitude, which have greatly influenced our research and provided us with many valuable ideas and suggestions. They have built a great foundation for the successful completion of this monograph.

We wish to thank the National Natural Science Foundation of China for providing financial support. Some of the research findings in the monograph come from a project of the National Natural Science Foundation of China, namely Research into Key Technologies of Emergency Response Decision Support System based on user-generated content (UGC) under Internet Environment (No. 71471008, 71420107025, 91224007). Thanks to the financial support we received, we were able to continue with our research.

Last, our thanks go to our beloved families for their loving consideration and great confidence in us all through these years. We also owe our sincere gratitude to our friends and fellow classmates who gave us their help and time in listening to us and helping us work out our problems during this difficult time.

Contents

Abstract

In recent years, a variety of natural and man-made disasters such as earthquakes, volcanoes, floods, hurricanes, chemical spills, nuclear leaks, epidemics, crashes, explosions, and urban fires have occurred frequently around the world. They are a direct cause of the loss of human lives and property and seriously damage the stability of our societies. Therefore, it is particularly important to be well prepared before an incident, to act in a timely fashion during an incident, and to cope with its aftermath. This is the reason for creating the Emergency Response Decision Support System (ERDSS).

This monograph first introduces what the ERDSS is and its commercial value. Based on an analysis of past events, it proposes some challenges and trends that may be encountered in the future. Chapter 2 is primarily concerned with an introduction to the ERDSS framework and ten related modules: Emergency Service Helpdesk, Command and Coordination Center, Emergency Plan Management, Emergency Relief Supplies Management, Emergency Finance Budget Management, Emergency Organization and Activity Management, Emergency Knowledge Warehouse, Emergency Warning Management, Emergency Alarm Management, and Problem Analysis and Management. Chapter 3 introduces key technologies applied in the ERDSS, which have great practical value in everyday life, especially Disaster Assessment, Adaptive Information Evaluation, Knowledge Management, Knowledge Evaluation, Blog Evaluation, Information Dissemination Analysis, Emergency Resource Scheduling, and Simulation Modeling. Chapter 4 provides two typical examples of practice to demonstrate the crucial role that ERDSS plays in the business processes of emergency incidents. Simultaneously, the two examples demonstrate the technologies and models described earlier in the monograph and illustrate their continuous improvement, high application, and promotional values. The first example is the Disaster Rescue Decision Support System for Beijing, which makes use of information technology and communications, simulation, system optimization, and emergency response decision support technologies, with a focus on emergency disaster rescue and security work in Beijing. This system optimizes Beijing's disaster relief and security system and establishes the decision support system for disaster rescue in Beijing. The system can help to position and store rescue resources before a disaster, provide effective support for coordination and command work during a disaster, schedule and distribute rescue resources after a disaster, and improve Beijing's disaster relief and security system's ability to deal with unexpected disasters and rescue work. The second example is Beijing's emergency flood management intelligent database system, which uses Apriori algorithms to mine the knowledge embedded in the Beijing flood emergency database. Knowledge representation, knowledge mining, and other key research technologies combine ontology, reasoning, and association rules to construct intelligent flood database systems. The Beijing system uses these key technologies and human–computer interaction as a subsystem of the decision support system, integrating human experience, and the rescue plan to support decision-making.

Keywords Emergency management · Emergency decision · Urban disasters · Key technology · Knowledge Management · User-generated content · Intelligent system · Disaster loss assessment · Adaptive evaluation

Chapter 1
Introduction

1.1 What Is Emergency Response

In recent years a variety of natural and manmade disasters such as earthquakes, volcanoes, floods, hurricanes, chemical spills, nuclear leaks, epidemics, crashes, explosions, and urban fires have occurred frequently around the world. They are a direct cause of the loss of human lives and property and seriously damage the stability of our societies.

The 2010 Haiti earthquake caused severe losses. The International Red Cross (2010) estimated that the quake affected about three million people; the Haitian Government (2011) reported that over 316,000 people were confirmed dead, 300,000 estimated injured, and 1,000,000 were made homeless.

A series of coordinated terrorist attacks aimed at several famous American buildings on September 11, 2001, caused major building fires, destroyed buildings in urban areas, and created serious damage to the surrounding environment that severely impeded the efficiency of emergency response.

Severe Acute Respiratory Syndrome (SARS), which spread to more than 25 countries after its initial outbreak in early 2003, claimed numerous lives and caused tremendous economic losses.

A stampede occurred on the Bund on December 31, 2014 that led to 36 people's deaths and 49 injuries. Tightly packed viewers and the commotion initiated by yet undetermined reasons contributed to the seriousness of this event. Although officers initiated evacuations and rescues as quickly as possible, many improvements are needed to minimize harmful consequences.

In January 2015, an acute blizzard affected Canada, the United States, and Europe, resulting in far more snow than usual in dozens of places, resulting in the cancellation of thousands of flights, suspension of normal activities, and some deaths in the worst affected areas, in spite of advance warnings and preparations.

At 23:34 on August 12, 2015, there was an explosion in a warehouse of the International Logistics Center Ruihai Company in Tianjin Port, where dangerous

© The Author(s) 2017
S. Shan and Q. Yan, *Emergency Response Decision Support System*,
SpringerBriefs in Business, DOI 10.1007/978-981-10-3542-5_1

goods were stored. The warehouse burned strongly, creating a mushroom cloud, and causing heavy casualties and property losses. The first explosion occurred at 23:34:06, approximating equal to a 2.3 magnitude earthquake or three tons of TNT. The second explosion occurred 30 s later, approximately equal to a 2.9 magnitude earthquake, or 21 tons of TNT. One hundred and sixty-five people were killed in the explosion (including 24 police firefighters involved in the rescue disposal on site, 75 firefighters from Tianjin Port, 11 police officers, and 55 employees or local residents). Eight people were missing (five Tianjin firefighters and three family members of local firefighters), 798 people were injured (58 with severe injuries and 740 with less serious injuries) and 304 buildings, 12,428 goods vehicles, and 7533 containers were damaged. As at December 10, 2015, the direct economic loss stands at 6.866 billion yuan, based on statistical standards for enterprise workers' economic losses and other statistical standards and regulations.

Emergency response is a series of organized and coordinated precautions and actions during the time between the detection of a possible event and stabilizing the situation (Drabek 1991). Rapid and appropriate emergency responses are needed urgently to help achieve more efficient rescue and relief to reduce disastrous losses. For instance, the quality of immediate action in the first precious search-and-rescue days after a terrible earthquake largely determines the eventual number of casualties (Friedrich et al. 2000). The primary goal of an emergency response is to ensure the safety of the public and emergency responders. In addition, public and private property and the environment should be protected.

Effective emergency preparedness for and response to major events requires coordinated planning and action by multiple players from multiple first responder disciplines, jurisdictions, and levels of government and nongovernmental entities.

There are four principal emergency response functions—emergency assessment, hazard operations, population protection, and incident management. The four functions provide a framework for organizing the activities involved in responding to a wide variety of emergencies, natural hazards, technological accidents, or deliberate terrorist attacks or sabotage.

1. Emergency assessment

Emergency assessment activities in the response phase are directed toward intelligence—understanding the behavior of the hazard agent and the people and property at risk.

2. Hazard operations

Hazard operations have the same purpose as hazard mitigation in emergency management, but they are implemented only when the need arises. Their applicability varies considerably from one hazard to another.

3. Population protection

Information collected during the emergency assessment function is the basis for choosing population protection actions. Emergency managers must oversee the technical and organizational mechanisms by which the community emergency response organization will protect its own personnel and the public.

4. Incident management

This function demands centralized planning for command and control across a variety of local public sector, private sector, and non-governmental organizations. Further, it requires a strategy for coordinating their collective responses, as well as specifying how extra-community resources will be mobilized and integrated into the response effort.

1.2 What Is a Decision Support System (DSS)?

"A Decision Support System is a class of highly sophisticated, computer-based information systems employed by executives, managers, or policy makers to serve specific functionalities in managing corporate finance, marketing, planning, and operations" (Power et al. 2002). Commonly, it integrates computer tools for data collection, analysis, and report, with decision models to support the organization's management. Typically, this occurs through gathering information from the organization's business processes and the marketplace to offer relatively abstract knowledge as the basis for timelier and better-considered decisions.

An organization's information processing facilities consist of three components for planning and analysis: central or distributed processing hardware, a database designed for high-volume transaction processing, and data (both external and that generated within an organization). These applications fall into the upper level of the DSS triangle.

If DSS is to make an impact on company management, it must meet these four requirements:

1. Provide management with a usable interface to a shared pool of summarized data from inside and outside the company, such as market competition and the economy;
2. Provide manipulative and analytical tools to transform the data in any way management requires;
3. Provide presentation facilities through the use of advanced graphics to show the results of analysis; and
4. provide modeling facilities to examine a range of scenarios.

The generally accepted view of a DSS is that it is an interactive computerized system consisting of three major components: a dialog subsystem, a database subsystem, and a model base subsystem (Watson and Sprague 1993); or an interface subsystem, a knowledge subsystem, and a problem processing subsystem (Holsapple and Whinston 1996). A DSS is designed to help a decision maker solve managerial decision problems interactively, using the knowledge and other capabilities embodied in its components. The three-component architecture is capable of managing data, fitting data into models, and providing methods to reach decisions (Angehrn and Jelassi 1994). By manipulating models and data, the decision maker is able to examine various scenarios and their consequences. The user interface component, which may be tailored to the user's individual preferences and expertise, is a user-friendly and effective communication facility. The three components together contribute to the quality of the decisions taken by the decision maker.

Currently, decision support systems help decision makers to access and understand data better and understand the implications of their judgments regarding that data, to assist them in making more educated decisions with the information available (French 2004). This brings numerous benefits, such as speeding up the decision-making process, increasing organizational control, and helping to automate managerial processes. A decision support system for emergency response and management, however, must be tailored to support operational, tactical, and strategic decision making-processes (French and Turoff 2007), able to help optimize task assignment and resource allocation and guide long-term decisions (Thompson et al. 2006).

1.3 What Is an Emergency Response Decision Support System (ERDSS)?

An emergency response decision support system is a decision support system that integrates all the special functionalities of emergency management and response into a system to take advantage of the possibilities that information technologies can provide for emergency response (discussed further below). After a disaster, for example, emergency planners are able to benefit from such systems to detect the spatial location of events, allocate shelters, emergency facilities, and resources to people at risk, and analyze transportation routes. An advanced ERDSS also allows managers to visualize immediate actions, such as the evacuation of the affected population from a disaster site.

However, numerous shortcomings in current emergency relief operations inhibit optimal decision making during disaster management operations, including accessibility and validity of information, lack of standardization, and disordered collaboration (NRC 1999; IAFF 2005). The 9/11 Commission Report (2004) reported that effective decision making during emergency response operations was hampered by problems in command and control and in internal communications to the extent that "incident commanders from responding agencies lacked knowledge of what other agencies and, in some cases, their own responders were doing" resulting in command, control, and communications problems. The report identified the need to "enable first responders to respond in a coordinated manner with the greatest possible awareness of the situation."

To provide adequate situational awareness and decision-making support to manage crisis situations, researchers and practitioners in disaster management have urged the development of emergency response decision support systems that can support first responders by enhancing their situational awareness and lead to better decision-making (Klann 2008). "First responders" are individuals who, in the early stages of an incident, are responsible for the protection and preservation of life, property, evidence, and the environment, as well as emergency management, public health, clinical care, and public works. The term includes other skilled support personnel that provide immediate support services during prevention, response, and recovery operations.

An Emergency Response Decision Support System must assist decision makers to evaluate emergency plans and select an appropriate plan of action by supporting heterogeneous emergency response data sources and providing decision makers with access to appropriate emergency rescue knowledge. It also needs to provide differentiated services to meet particular requirements.

An Emergency Response Decision Support System must work in an extreme and stress-filled environment, accessing static information such as road maps and building floor plans as well as dynamic and real time information such as the latest disaster developments and the current location of emergency personnel and resources. As an emergency evolves, requirements (both informational and logistical) may change, resulting in necessary modifications to the response (Wang et al. 2008, 2009). An investigation of first responders' requirements in a Dutch emergency response demonstrated that much of the information first responders request during a crisis is dynamic and needed almost instantaneously (Diehl et al. 2006). Furthermore, a desirable ERIS platform consists of a number of Mobile Data Terminals (MDTs) (Lindsay et al. 2009), and many handheld devices such as mobile phones, iPads, personal digital devices (PDA), connected to one or more large-scale computer server systems located in a fixed place. These features make it desirable for an emergency response decision support system to be global and possess distributed information systems capable of real time information acquisition, processing, sharing, and analyzing.

1.4   ERDSS Business Values

Great profits and savings can be made by making use of emergency response decision support systems. Deqia et al. (2012) argues that an emergency response decision support system for highway traffic accidents can minimize average response times in different types of accidents. Quinn and Jacobs (2007) insisted that an emergency response management system could assist in forecasting episodes and guiding remedial measures to prevent damage from future events. Yoon et al. (2008) proposed an emergency response decision support system to verify a transportation agency's capabilities, procedures, and preparedness to handle a simulated emergency involving a combination of terrorist attacks, natural hazards, and severe traffic accidents. It also enabled observation and self-examination of strengths and weaknesses in managing, preparing, and responding to a series of security emergency scenarios. Zhen Liang Liao (2012) explored whether such a system can generate emergency plans quickly and accurately to meet the requirements of an emergency on-site.

The main benefits (Yoon et al. 2008) of emergency response decision support systems result from creating more effective and efficient operations and responsive systems based on real-time information sharing and decision-making. Access to a decision support system provides a decision maker with the ability to (1) identify, secure, and deploy the correct number and types of resources in real time; (2) determine the current inventory of available resources, personnel, and their location; (3) review historical records of similar or related events; and (4) store on-time decisions for future review and create an organizational memory system. In a highly distributed organization such as a state department of transportation, a decision support system functions as a data warehouse to serve all the functions discussed above. Furthermore, it also enables all members of the distributed network to access information at any time and at any location in the state.

1.5 Current ERDSS Research

In recent years, significant effort has been devoted to providing computer support for emergency response. Geographical Information System (GIS)-based applications have been developed to help decision makers to analyze, manage, and respond to emergencies by situating incident information in its geospatial context. Specific examples include emergency management systems for containing chemical and nuclear pollutants, monitoring the risk of oil pollution, and tracking and visualizing the predicted course of hurricanes. An Artificial Intelligence (AI)-based emergency response system has been used for environmental monitoring, and an example of the use of a knowledge-based model for decision support during flood emergencies can be found in example. Multi-agent systems have been used in large, complicated environmental emergency systems.

The ever-changing nature of today's world has switched the focus of responding to emergencies; "new realities are now making strategy itself appear obsolete, turning businesses into adaptive systems that remain alert to shifting paradigms and play-out different scenarios in a sense-and-respond mode" (Lin et al. 2004 6).

1.6 ERDSS Challenges and Trends

1. ERDSS Challenges

(1) Equipment is often tied to victims

Biosensors wired to a monitor are a predominant part of existing biosensor equipment used in emergency response. The use of such equipment is severely hampered by the fact that data can be seen only when standing next to the monitor, which itself must be next to the victim. Furthermore, it is a cumbersome task to move a victim from one location to another when biosensors attached to the victim are wired to a monitor.

(2) Identification of victims is difficult

It is extremely difficult to obtain and maintain valid and credible identity of victims in emergencies. The identity of a person can be established credibly only when the person is conscious or by third part identification.

(3) Situational overviews are very incomplete and primarily in the heads of the professionals involved

During emergency response, it is very difficult for the professionals involved to build and maintain an overview of an incident site and the rescue effort. Maps and sketches are on paper and co-located people only can share them. Indeed, most overview information distributed among people exists only in their heads. Hence, it is difficult to create a common understanding of the incident scene: the incident as such, injured and uninjured persons, and available human and other resources. It is difficult to plan the use of the area surrounding the incident, access routes for ambulances can be difficult to describe and communicate, danger areas hard to locate and avoid, etc.

(4) Communication on site is primarily face-to face

Currently Information and Communication Technology (ICT) support for communication involving an incident site is restricted essentially to two-way radios that have a few additional push-button commands. Typically, each professional group (police, ambulance staff, medical staff, and fire fighters) uses their own radio frequency, which prevents different professional groups from communicating with each other. In principle, the managers of each group have a separate set of radios for their communication, but in practice they usually have to find each other physically and try to stay together to communicate and coordinate action.

(5) Equipment and systems change with every situation and as specific situations unfold

This challenge has two dimensions. The first concern changes in systems over time, including the integration of new technological components. The second concerns dynamic changes to systems and equipment during emergency response. Emergency response personnel use several different systems and devices including radios, medical equipment, GPS-equipped vehicles, call alarm and dispatch systems, and health records. These devices and systems are not based on a common high-level design vision. Typically, they are poorly integrated and different elements evolve, become operational, and are discarded largely independent of other elements, which may create problems with "neighboring" elements.

As outlined above, emergency response involves many different people, devices, and systems. It is a challenging task during emergency response to facilitate the dynamic changes they are subject to. As an illustration, radio communication between professionals at the accident site and other locations (hospital or police station) is often difficult. Mobile phones are used in some cases, but in major incidents mobile phone networks frequently become overloaded and break down.

(6) Equipment and systems vary considerably with respect to reliability and trustworthiness

Emergency response is often a matter of life and death; therefore, the demand for reliability of the most critical patient-related devices is especially high. Non-patient related devices and systems such as radios and communication between radios and call alarm systems vary in quality; several are of poor quality. As an illustration, radio communications between the incident site and coordination and control centers are often error-prone due to poor radio network coverage. At the bottom of the list for reliability and trustworthiness are some systems that seek to automate emergency vehicle dispatch. The character of the coordination activities in dispatch centers in emergencies makes it very difficult to develop reliable systems that automate this work.

(7) Suitability and immediate usability determines what equipment and systems are actually used

Two problems recur in the use of existing support for emergency response: the support is not suited to real-life emergency response and/or the professionals are not familiar with specific devices and systems. They have received training in their use on special courses but, in the hectic unfolding of an incident, professionals often fail to use equipment efficiently, stop using it altogether, or do not use it at all. This is simply because the equipment has not played a role in their daily routine, or because it was not suited to the job in the first place. As an illustration, professionals are trained to use special incident cards for major incidents, but they do not use the cards in real incidents. One major reason is that it takes too much time to use them. Special radios for major incidents are sometimes forgotten because they are not part of daily routines.

2. ERDSS trends

(1) Statistical technology is becoming widely used

The frequency of all kinds of accidents appears to be increasing, at least at a high linear rate, and associated costs may be increasing at an exponential rate. This is a very serious trend and demands immediate attention. Thus, statistics provide a means to evaluate our capability to control the consequences of emergencies, to determine classes of emergencies that need attention, to decide whether we are spending an appropriate amount to handle disasters, and to predict where and when disasters are likely to occur. Clearly, we need a more organized and standardized method of collecting and reporting emergency statistics in a manner similar to that used in crime control. The value of such statistics is tremendous.

(2) Coordination in emergency response will become more important

Effective coordination is an essential ingredient in emergency response management. Coordination is demanding as it involves conditions typical of an emergency such as, for example, high uncertainty, the need for rapid decision making, and response under temporal and resource constraints.

Emergency response coordination is complex: it involves factoring in exigencies such as great uncertainty, sudden and unexpected events, the risk of possible mass casualties, significant time pressure, severe resource shortages, large-scale impacts and damage, and disruption of necessary infrastructure such as electricity, telecommunications, and transportation. This is complicated by infrastructure interdependencies, multi-authority and massive personal involvement, conflict of interest, and high demand for timely information.

References

9/11 Commission Report (2004) Final report of the National Commission on terrorist attacks upon the United States—executive summary. Available at www.9-11commission.gov/report/index. htm. Accessed 7 May 2008

Angehrn A, Jelassi T (1994) DSS research and practice in perspective. Decis Support Syst 12:267–275

Deqia H, Xiumina C, Zhea M (2012) A simulation framework for emergency response of highway traffic accident. Proc Eng 29:1075–1080

Diehl S, Neuvel JMM, Zlatanova S, Scholten HJ (2006) Investigation of user requirements in the emergency response sector: the Dutch case. In: Second symposium on Gi4DM, Goa, India, CD ROM, 6 pp

Drabek TE (1991) Introduction. In: Drabek TE, Hoetmer GJ (eds) Emergency management: principles and practice for local government. International City Management Association, Washington DC, pp xvii–xxxiv

French S (2004) Decision analysis and decision support. Wiley, New York

French S, Turoff M (2007) Decision support systems. Commun ACM 50(3)

Friedrich F, Gehbauer F, Rickers U (2000) Optimized resource allocation for emergency response after earthquake disasters. Saf Sci 35(1):41–57

Haitians recall 2010 quake "hell" as death toll raised. Many survivors of the earthquake were left with permanent disabilities including limb amputation, spinal cord injury and severe fractures. Traditionally disability was not well received in Haiti. Team Zaryen, a Haitian Amputee Soccer Team has been challenging that negative association by showing their success on the pitch. Rueters, 2011-01-12

Holsapple CW, Whinston AB (1996) Decision support systems: a knowledge-based approach. West Publishing, St. Paul

IAFF (2005) Fire fighters doing their jobs and the jobs of others. Available at www.iaff.org/ Comm/Katrina/Center0903Press.asp. Accessed 7 May 2008

Klann M (2008) Tactical navigation support for firefighters: the LifeNet ad-hoc sensor network and wearable system. In: Proceedings of the second international workshop on mobile information technology for emergency response (Mobile Response 2008), Bonn, Germany, pp 41–56

Lin GY, Luby Jr RE, Wang KY (2004) New model for military operations. ORMS 31(6)

Lindsay R, Cooke L, Jackson T (2009) The impact of mobile technology on a UK police force and their knowledge sharing. J Inform Knowl Manag 8(2):101–112

National Research Council (NRC) (1999) Reducing disaster losses through better information. National Academic Press, Washington, DC

Power DJ, Sharda R, Burstein F (2002) Decision support systems. Wiley, New York

Red Cross: 3M Haitians Affected by Quake (2010) CBS News. 13 Jan 2010. http://www.cbsnews. com/stories/2010/01/13/world/main6090601.shtml?tag=cbsnewsSectionContent.4. Retrieved 13 Jan 2010

Quinn NWT, Jacobs KC (2007) Design and implementation of an emergency environmental response system to protect migrating salmon in the lower San Joaquin River, California. Environ Model Softw 22:416–422

Thompson S, Altay N, Green III WG, Lepetina J (2006) Improving disaster response efforts with decision support systems. Int J Emerg Manag 3(4)

Wang J, Rosca D, Tepfenhart W, Milewski A, Stoute M (2008) Dynamic workflow modeling and analysis in incident command systems. IEEE Trans Syst Man Cybern Part A Syst Hum 38 (5):1041–1055

Wang J, Tepfenhart W, Rosca D (2009) Emergency response workflow resource requirements modeling and analysis. IEEE Trans Syst Man Cybern Part C Appl Rev 39(3):1–14

Watson HJ, Spragure RH Jr (1993) The components of an architecture for DSS. In: Sprague RH Jr, Watson HJ (eds) Decision support systems: putting theory into practice, 3rd edn. Prentice-Hall, Englewood Cliffs

Yoon SW, Velasquez JD, Partridge BK, Nof SY (2008) Transportation security decision support system for emergency response: a training prototype. Decis Support Syst 46:139–148

Chapter 2
The Emergency Response Decision Support System Framework

An emergency response decision support system (ERDSS) needs to assist decision makers to evaluate emergency plans and select an appropriate plan of action during an emergency by supporting heterogeneous emergency response data sources and providing decision makers with access to appropriate emergency rescue knowledge. It also needs to provide differentiated services to meet particular requirements. This chapter describes the ERDSS framework and its main components in detail.

2.1 The ERDSS Framework

We propose an ERDSS framework that consists of ten functional modules: Emergency Service Helpdesk, Command and Coordination Center, Emergency Plan Management, Emergency Relief Supplies Management, Emergency Finance Budget Management, Emergency Organization and Activity Management, Emergency Knowledge Warehouse, Emergency Alarm Management, and Problem Analysis and Management. The ERDSS framework shown in Fig. 2.1 describes the main components of the system and their relationships.

For example, the scope of ontology modeling in flood emergency databases is based strictly on the principle of the smallest ontology agreement, and knowledge about the relationship between flood and flood emergency rescue is expressed through ontology. Flood rescue ontology consists of two parts; flood description and rescue plan description. The two parts include different levels of ontological concepts and form the hierarchy of the flood ontology description system.

© The Author(s) 2017
S. Shan and Q. Yan, *Emergency Response Decision Support System*,
SpringerBriefs in Business, DOI 10.1007/978-981-10-3542-5_2

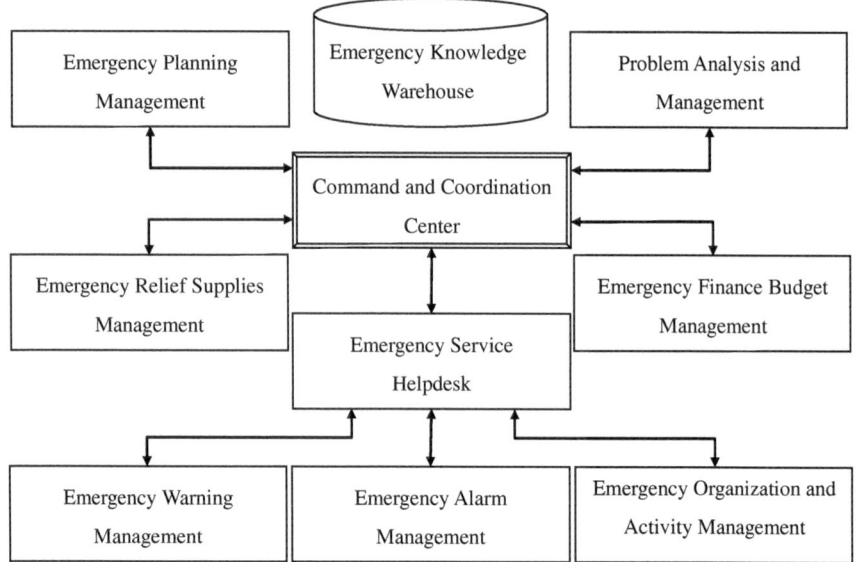

Fig. 2.1 The ERDSS Framework

2.2 The Emergency Service Helpdesk Module

Decision making in emergency response is an extremely time-sensitive and challenging task that requires immediate and effective response from decision makers who are surrounded by a variety of uncertain information and are under huge pressure from the need to coordinate action (Green and Kolesar 2004). The Service Desk is a key tool to collect, share and disseminate communications and emergency information efficiently and undertake preliminary analysis. These roles are important in helping decision makers make timely and effective decisions during emergency response. However, Service Desk is not an easy role, because it deals with diversified external information and coordinates various organizations.

During emergency response, the ERDSS will receive a large number of requests for timely information screening and processing. Therefore, the ERDSS interface needs to integrate multiple data sources and communication channels at a simple, single contact point that collects emergency information from the public and maintains good communications with the public and external organizations. The Service Helpdesk, a module that connects public emergency information sources and other modules such as the Command and Coordination Module, is the only information interface in ERDSS that collects, analyzes, and releases emergency information. The module processes existing public emergency information and provides formal emergency response progress reports to the public. It also communicates with other emergency organizations.

The Emergency Service Helpdesk Module provides the following services:

1. Acts as a window for contact information collection and distribution;
2. Provides preliminary emergency data processing including analysis, normalization, and filtering;
3. Acts as the sole communication channel with other organizations;
4. Acts as the sole emergency information interface with other modules; and
5. Keeps the system scalable and flexible.

The Emergency Service Helpdesk Module corresponds to Service Desk in the Information Technology Infrastructure Library (ITIL). The module's architecture is shown in Fig. 2.2.

The Emergency Service Helpdesk Module (see Fig. 2.2) has three components: the Distributed Heterogeneous Contact Information Interface, Preprocessor, and the Unified Data Interface. The Distributed Heterogeneous Contact Information Interface is an entry point collecting data about emergencies from the public, a window releasing emergency reports to the public, and a channel communicating with other organizations.

Preprocessor functions are preliminary analysis, information classification and normalization, communication process design, and feedback integration. The

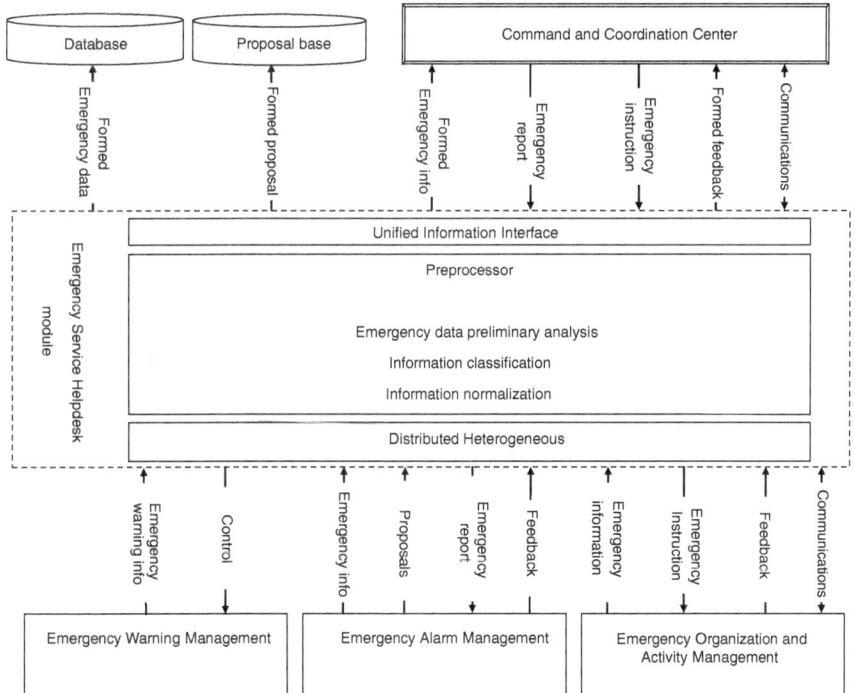

Fig. 2.2 Architecture of the emergency service helpdesk module

Unified Data Interface is an easy-to-use programming interface that offers formatted information and allows other modules to convert and exchange the data they need.

It is reasonable in government emergency response decision-making to use blog resources to help build a demand-oriented self-adaptive evaluation model. The subject is the information demand unit of emergency response decision-making, and the priority influence factor is the demand tendency of decision-making for information.

2.3 The Command and Coordination Center Module

Decision makers face three main challenges: how to propose reasonable emergency rescue plans, how to issue clear emergency commands, and how to coordinate the various organizations involved in emergency responses. "Command" refers to the way decision-makers organize and lead emergency responses. "Coordination" means correctly handling various internal and external relations among emergency organizations and promoting the goals of emergency rescue by creating favorable conditions for emergency rescue.

From the perspective of management, emergency coordination is a managerial function that organizes and coordinates different rescue activities during emergency response. Command and coordination are challenging because they involve multiple organizations and deal with uncertain and time-sensitive information.

An ERDSS needs to integrate emergency decision-making measures and coordinate multiple organizational objectives to identify the severity of an emergency, select and implement appropriate plans, make adjustments according to the actual situation, and monitor the results after response plans are implemented.

Therefore, the Command and Coordination module, located at the core of an ERDSS, is required to work as a command center to issue emergency instructions and as a coordination center to enable emergency organizations to work together to achieve shared rescue goals.

The Command and Coordination Center Module primarily handles existing emergency work. It may also run analyses to identify the causes of an emergency so that similar disasters do not reoccur.

The Command and Coordination Center Module provides the following services:

1. Identifies and confirms the severity of an emergency;
2. Monitors the entire emergency response process;
3. Makes emergency response decisions;
4. Coordinates multiple organizations to conduct rescue work;
5. Assesses the level of victims' satisfaction; and
6. Modifies timely rescue measures.

Fig. 2.3 Architecture of the command and coordination center module

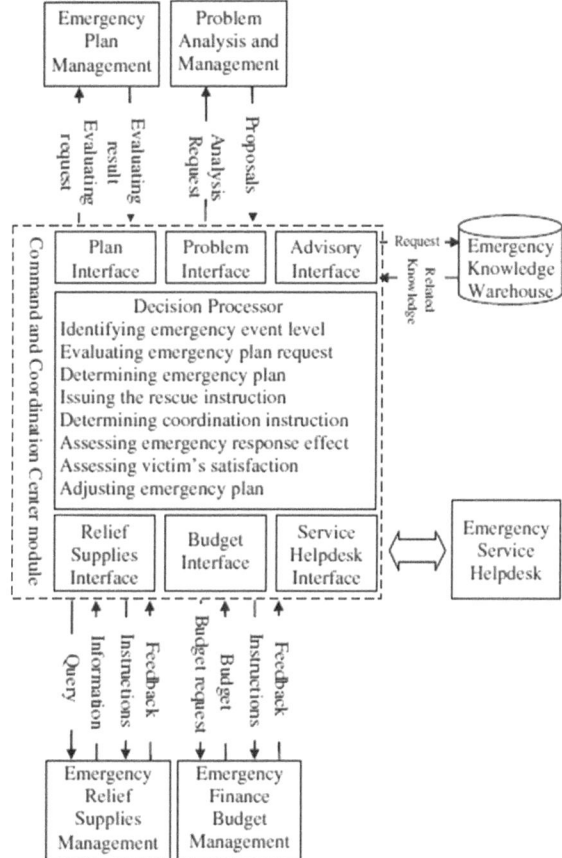

The Command and Coordination Center Module corresponds to the processes of Incident Management, Service Level Management, and IT Service Continuity Management in ITIL. The model's architecture is shown in Fig. 2.3.

The Command and Coordination Center Module (see Fig. 2.3) has seven components: Plan Interface, Advisory Interface, Problem Interface, Decision Processor, Relief Supplies Interface, Service Helpdesk Interface, and Budget Interface. Plan Interface's functions are evaluating the entry of emergency plans, submitting rescue plans for assessment, and receiving assessment results. Advisory Interface is a knowledge base that emergency decision makers can access which allows them to obtain emergency response knowledge such as contingency plans, emergency laws, emergency response rules, typical cases, emergency expert information, and other related knowledge. Problem Interface undertakes in-depth analysis to identify the causes of emergencies and measures to deal with them. Decision Processor identifies the scale of emergencies, determines which emergency plan should be chosen; issues rescue instructions, coordinates rescue operations, and assesses emergency response results. Relief Supplies Interface offers relief supplies, information, and feedback.

Service Helpdesk Interface handles information from the Emergency Service Helpdesk Module. Budget Interface is a channel for finance budget requests channel and handles applications for emergency relief funds.

Based on the analysis of the User Generated Content (UGC) characteristics of urban disasters, we are able to design a UGC-based urban disaster loss assessment index, build a dynamic and continuous monitoring and analysis algorithm for disaster, and study the disaster loss incremental assessment process to support disaster emergency response decisions.

2.4 The Emergency Plan Management Module

The importance of emergency plans is obvious when an emergency occurs that requires efficient emergency procedures (Ingram 2010). An emergency plan is a group of procedures that are implemented during an emergency that involves communication, planning, action, risk analysis, operational support, logistic support, and whatever is necessary to reduce impacts (Calixto and Larouvere 2010). Emergency plans seek the most efficient way to use necessary resources to meet urgent needs under conditions of emergency (Alexander 2005).

Emergency plans are major challenges for governments in many countries because there are different opinions about the appropriate emergency framework (Calixto and Larouvere 2010). Although every emergency is unique, they share the same response process: forecasting, making predictions, issuing warnings, predicting consequences, and developing plans. An emergency plan should consider an emergency from multiple angles. The ability of an emergency plan to deal with an emergency depends on its completeness, operability, effectiveness, flexibility, rapidity, and rationality (Cheng and Qian 2010). Alexander (2005) proposed 18 principles to judge the quality of an emergency plan and presented a standard for local government emergency plans.

The development and management of an emergency plan relies on a research center to study emergency requirements and risk, a development center to draw up and test the emergency plan, and a supervision component to supervise the plan's use and revision. The Emergency Plan Management Module located at the core of the ERDSS meets these requirements.

The Emergency Plan Management Module provides the following services:

1. Runs emergency requirements analysis and risk assessment;
2. Supports emergency classification;
3. Maintains the algorithms for evaluating emergency plan effectiveness;
4. Supports emergency plan modeling;
5. Simulates the environment through optimization analysis;
6. Supports drawing up and testing emergency plans; and
7. Supports supervision of the use and revision of emergency plans.

Fig. 2.4 Architecture of the emergency plan management module

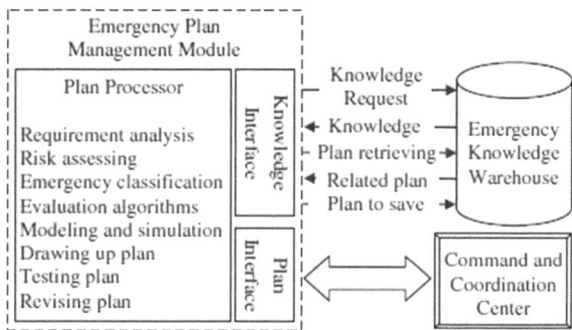

The Emergency Plan Management Module corresponds to Service Level Management, IT Service Continuity Management, Capacity Management, and Availability Management processes in ITIL. The model's architecture is shown in Fig. 2.4.

The Emergency Plan Management Module (see Fig. 2.4) has three components: Plan Interface, Knowledge Interface, and Plan Processor. Plan Interface handles the interactive information between this module and the Command and Coordination Center Module. Knowledge Interface is a channel retrieving and using emergency knowledge such as cases, rules, laws and regulations, data, models, and related emergency plans. Plan Processor functions include analyzing emergency requirements, assessing emergency risk, classifying emergency events according to severity and scope of impact, evaluating emergency plans according to the evaluation index system and algorithms, modeling emergency plans and simulation analysis, drawing up and testing specified emergency plans, releasing emergency plans stored in the knowledge warehouse, and revising outdated or defective plans.

In the case of the "7.21" heavy rainstorm in Beijing, the first action taken was to collect all relevant disaster data and extract the time parameters needed for the model. Second, macro and micro diagnostic analysis and evaluation were carried out in accordance with the model evaluation system. Third, CPN (CCSDS Principle Net) simulation tools were used for simulation experiments. Finally, policy recommendations were made based on the results of performance analysis and simulation results.

2.5 The Emergency Relief Supplies Management Module

Relief Supplies Management is a type of emergency response and of social management. It serves broad social objectives rather than the benefit of an individual organization. Relief supplies operations rely heavily on logistics under uncertain, risky, and urgent situations. Therefore, supply chain management principles are applied in a different way from their application in traditional businesses.

Coordination, vehicle routing, and supply allocation decisions are critically important for relief supplies operations.

To deliver the right relief supplies in a timely manner to the right locations, an ERDSS relies on a research center to study relief supplies requirements and optimize routes, a schedule center to coordinate transportation in the relief supply chains, and a measurement point to obtain feedback and undertake victims' satisfaction analysis. The Emergency Relief Supplies Management Module is located at the core of an ERDSS.

The Emergency Relief Supplies Management Module provides the following services:

1. Determines categories and quantities of relief supplies;
2. Provides routing algorithms, modeling, and simulation;
3. Provides logistics operation and coordination;
4. Manages relief supplies distribution;
5. Collects relief supplies feedback and undertakes victims' satisfaction analysis; and
6. Provides instructions on how to execute emergency rescues.

The Emergency Relief Supplies Management Module corresponds to Service Level Management and Capacity Management Processes in ITIL. The model's architecture is shown in Fig. 2.5.

The Emergency Relief Supplies Management Module (see Fig. 2.5) has three components: Relief Supplies Interface, Knowledge Interface, and Relief Supplies Processor. Relief Supplies Interface handles the interactive information about relief supplies between this module and the Command and Coordination Center Module. Knowledge Interface is a channel to retrieve and use emergency knowledge related to relief supplies. Relief Supplies processor functions are determining categories and quantities of emergency relief supplies, providing inventory management and

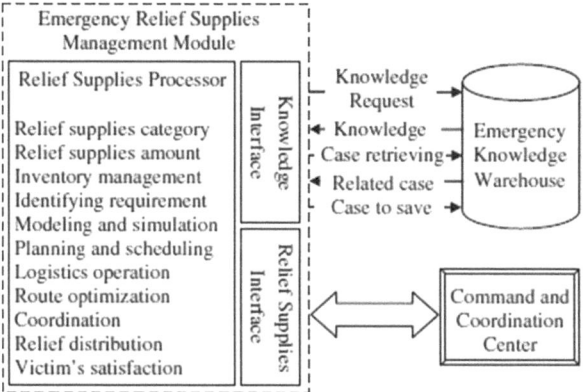

Fig. 2.5 Architecture of the emergency relief supplies management module

statistics reports, identifying relief supplies requirements, optimizing routes, modeling and simulation, coordinating relief supplies logistics, distributing relief supplies, and running victims' satisfaction analysis.

In the case of the "7.21" heavy rainstorm in Beijing, it was necessary to combine the emergency plan for an urban natural disaster and the plan for rational use of the emergency materials reserves, and to establish a unified material management platform based on GIS. The platform interacts seamlessly with the rescue decision support platform. Further, it achieves unified scheduling of emergency supplies management in the event of natural disasters.

2.6 The Emergency Finance Budget Management Module

The cost of emergency response is enormous. The more severe an emergency, the greater the emergency response costs. Emergency relief activities consume a significant amount of labor, materials, and financial resources. Costs must be monitored to ensure the quality of emergency rescue. Barfod et al. proposed a concept of composite decision support for complex decisions that combines cost-benefit analysis, multi-criteria decision analysis for economic assessment, and strategic impact assessment (Barfod et al. 2011).

To balance rescue costs and the rescue effect, an ERDSS needs a budget center to provide sufficient funds for emergency operations according to emergency plans, an accounting center to monitor the cost of emergency activities, and a finance report center to analyze and evaluate the effect of emergency activities from a cost perspective. The Emergency Finance Budget Management Module located at the core of an ERDSS meets these requirements.

The Emergency Finance Budget Management Module provides the following services:

1. Provides financial plans consistent with emergency plans;
2. Provides emergency accounting;
3. Monitors the cost of emergency activities;
4. Supports budget models and provides emergency cost prediction algorithms;
5. Chooses emergency cost structures and elements; and
6. Provides emergency financial reports for in-depth analyses.

The Emergency Finance Budget Management Module corresponds to the IT Service Finance Management Process in ITIL. The model's architecture is shown in Fig. 2.6.

The Emergency Finance Budget Management Module (see Fig. 2.6) has three components: Finance Budget Interface, Knowledge Interface, and Finance Budget Processor. Finance Budget Interface handles the interactive information about emergency finances and budgets between this module and the Command and Coordination Center Module. Knowledge Interface retrieves the knowledge related

Fig. 2.6 Architecture of the emergency finance budget management module

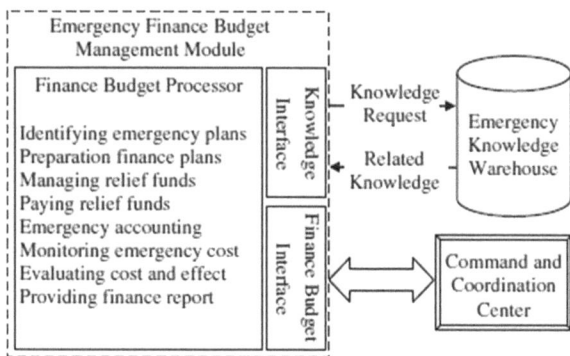

to finance budget and accounting. Finance Budget Processor functions are identifying approved emergency plans, proposing budget targets, preparing financial plans, managing and paying relief funds, carrying out cost accounting, monitoring emergency operation costs, evaluating the balance of cost and effectiveness, and providing financial reports to emergency decision makers.

2.7 The Emergency Organization and Activity Management Module

The nature of emergencies, especially mass emergencies, means that emergency management teams usually have to make decisions in stressful situations, full of ambiguous information, information overload, and a high level of uncertainty. This requires knowledge-based non-routine problem solving skills (Schaafstal et al. 2001). Emergency management requires teamwork across many teams from different organizations with different goals and cultures that are working together to minimize damage and loss from the emergency.

Emergency management requires good coordination and communication within and across teams. Not all emergency teams are effective or efficient when performing emergency activities (Siassakos et al. 2011). Effective performance evaluation of emergency response activities can help emergency organizations to identify missing relief activities and improve the efficiency of their emergency operations.

To manage organizations and emergency activities, an ERDSS needs an emergency plan execution system to undertake rescue activities and center for emergency scene information collection to provide feedback to the emergency command and coordination center. The Emergency Organization and Activity Management Module meets these requirements.

The Emergency Organization and Activity Management Module provides the following services:

1. Manages organizations, personnel, and rescue equipment;
2. Manages and monitors emergency rescue activities;
3. Supports rescue performance evaluation;
4. Executes emergency plans; and
5. Collects emergency scene information.

The Emergency Organization and Activity Management Module corresponds to the Incident Management process in ITIL. The module's architecture is shown in Fig. 2.7.

The Emergency Organization and Activity Management Module (see Fig. 2.7) has six functional components: Organization Management, Activity Management, Performance Evaluation, Activity Analysis, Knowledge Interface, and Organization Activity Interface. Organization Management manages various emergency organizations, personnel, rescue equipment, and facilities. Activity Management manages emergency rescue tasks, assignments, activities, training, communications, and

Fig. 2.7 Architecture of the emergency organization and activity management module

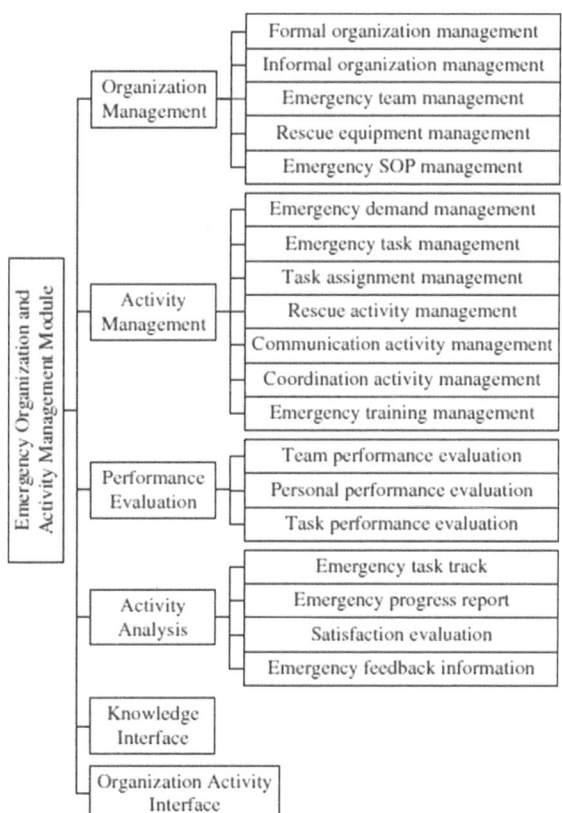

coordination. As one type of activity management, communication activity management handles formal and informal contacts, including phone calls, text messages, meetings, and so on. Coordination Activities Management resolves conflicts in rescue processes.

Performance Evaluation runs cost-impact analysis and other processes to improve the performance of emergency organizations and activities. Activity Analysis monitors the progress of emergency rescue activities, assesses emergency plans, and provides timely feedback to the Emergency Command and Coordination Center.

Knowledge Interface retrieves relevant knowledge from the knowledge warehouse. Organization Activity Interface deals with interactive information between this module and the Command and Coordination Center Module.

Blogs now have a low threshold and a very large internet presence, so they can assist us to collect information; however, emergencies are often powerful and destructive, making it increasingly difficult to locate valuable blogs. ERDMSS can solve this difficulty by clustering analysis and text analysis. We can identify target blogs with an emergency response theme for emergent events by combining a vector space model with a single scan clustering algorithm.

2.8   The Emergency Knowledge Warehouse Module

Knowledge can increase our capacity to take effective action. Emergency response knowledge can help emergency decision makers and managers to resolve problems effectively that they have not encountered previously. An emergency response system architecture embeds the knowledge required to support more effective emergency decision-making under different scenarios (Hernandez and Serrano 2001). What kind of knowledge is useful in the emergency response process? How can decision makers apply knowledge better in emergency decision-making and rescue activities? These emergency response problems need to be solved.

To make effective use of emergency knowledge, an ERDSS needs a warehouse to store different types of emergency knowledge and a decision support tool to support the emergency response. The Emergency Knowledge Warehouse Module meets these requirements.

The Emergency Knowledge Warehouse Module provides the following services:

1. Provides a warehouse for organizing and storing knowledge;
2. Offers emergency knowledge and analytical tools to support emergency activities;
3. Provides users with knowledge and interfaces to acquire knowledge from outside the system;
4. Enhances system intelligence; and
5. Improves emergency response capabilities.

Fig. 2.8 Architecture of the emergency knowledge warehouse module

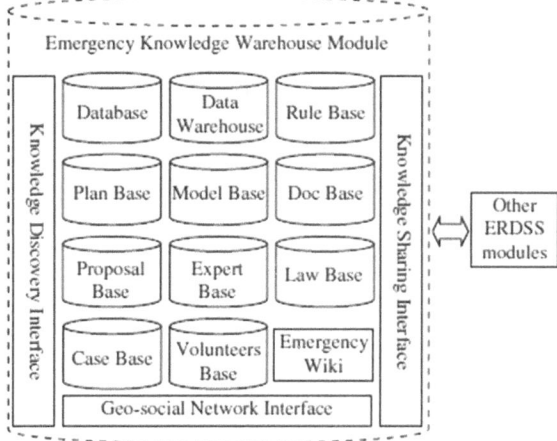

The Emergency Knowledge Warehouse Module corresponds to the Capacity Management Process in ITIL. The model's architecture is shown in Fig. 2.8.

The Emergency Knowledge Warehouse Module (see Fig. 2.8) has 15 functional components. The database stores current emergency response transaction data. The Data Warehouse stores historical emergency data to support data mining and online analytical processing (OLAP). Plan Base receives prepared emergency plans from the Emergency Plan Management Module and sends appropriate plans to the Command and Coordination Center Module.

Proposal Base stores formatted proposals from the Emergency Service Helpdesk Module to conduct further analyses. Model Base saves models and algorithms, such as plan analysis, route optimization, rescue budget, and simulation models. Case Base stores complete descriptions of typical emergencies. Rule base store structured emergency knowledge and experience. Doc Base stores documented knowledge such as reports, videos, and audios. Law Base stores laws and regulations related to emergency response and reflects the characteristics of e-government. Expert Base and Volunteer Base store information about emergency experts and volunteers respectively. Emergency Wiki is a wiki for emergency response purposes. A wiki is a type of website that allows users to add, modify, and delete its content via a web browser by using a simplified markup language or a rich-text editor (Lykourentzou et al. 2012). Knowledge Discovery Interface provides knowledge discovery tools and captures or collects emergency knowledge. Geo-Social Network Interface manages volunteer communities and captures the geo-positions of community members by connecting existing or new geo-social networks via specific APIs (Application Programming Interface). Knowledge Sharing Interface provides knowledge to support other modules in the ERDSS.

In case of flood emergency response, the case history database is collected to form a flood case and a case study is used to retrieve similar cases. The case description is based on ontology to provide a structured description framework for case knowledge. It provides an effective tool for knowledge management of historical cases. The system can use the current case automatically to generate the rescue plan based on historical cases and knowledge-mining rules, accelerate the formation of the rescue plan, and improve the organization's emergency management ability.

2.9 The Emergency Warning Management Module

Emergency warning or forecasting is an important function of emergency responses. Emergency warning or forecasting can help decision makers to take emergency measures and reduce disaster losses as early as possible. An ERDSS needs forecasting facilities to predict possible emergencies. The Emergency Warning Management Module is designed for emergency prediction.

The Emergency Warning Management Module provides the following services:

1. Collects safety information;
2. Supports prediction algorithms and data analysis;
3. Provides security status threshold settings; and
4. Manages related facilities and personnel.

The Emergency Warning Management Module corresponds to the Service Desk process in ITIL. The model's architecture is shown in Fig. 2.9.

The Emergency Warning Management Module (see Fig. 2.9) has two functional components: the Warning Management Processor and the Warning Interface. The Warning Management Processor collects and analyzes safety information and sends warning information to the Emergency Service Helpdesk Module. The safety information collection component inside the Warning Management Processor

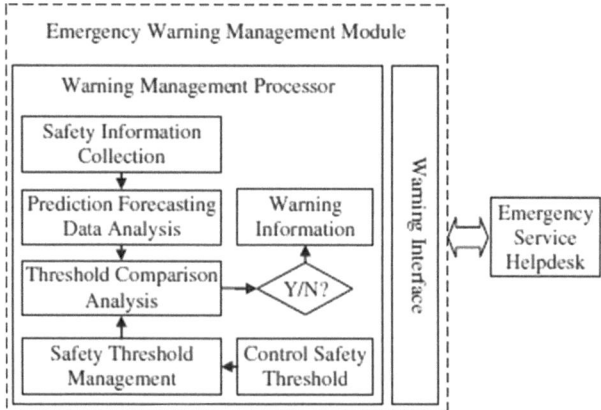

Fig. 2.9 Architecture of the emergency warning management module

gathers relevant data in major protection areas on a regularly and irregular basis. Safety information is obtained from personnel and purpose-built facilities. The emergency prediction and forecasting functional component analyzes scattered data gathered by the safety information collection functional component. The data threshold for security status is responsible for security alerts and controls various safety standards in the warning process. Warning Interface handles the interactive information between this module and Emergency Service Helpdesk Module.

A 6.5 magnitude earthquake occurred at 16:30 on August 3, 2014, in Zhaotong Ludian County, Yunnan. In the face of such an emergency, forecasting the demand for materials, raising, scheduling, and monitoring materials, and feedback links rapidly generated the need for up-to-date information. That is to say, these activities are prone to stagnation. The first reaction after the accident is very important. The command and control group needs to have demand forecast experts and to construct the Petri net model of the scheduling process for urban emergency supplies. With these resources, the group can rapidly identify the right kinds of goods and forecast the approximate number, thereby reducing the delay caused by uncertain resource allocation decisions.

2.10 The Emergency Alarm Management Module

Emergency alarms report emergency information to related organizations and departments when emergencies occur. Emergency alarms should involve both government departments and the public. The Emergency Alarm Management Module in an ERDSS collects emergency and demand information and provides emergency proposals and feedback to the Service Helpdesk Module. Lu and Yang (2011) proposed a hierarchical model to examine the mechanism of exchanging information about natural disaster responses in virtual communities.

The Emergency Alarm Management Module provides the following services:

1. Collects emergency information;
2. Collects emergency rescue demands;
3. Collects emergency response feedback;
4. Provides emergency rescue proposals submitted by the public; and
5. Connects external information.

The Emergency Alarm Management Module corresponds to the Service Desk process in ITIL. The model's architecture is shown in Fig. 2.10.

The Emergency Alarm Management Module (see Fig. 2.10) has four functional components: Dedicated Alarm Facilities, Emergency Call Center, Internet Emergency Information Center, and Alarm Interface. Dedicated Alarm Facilities detect and transmit emergency information to the Service Helpdesk Module. The facilities include special equipment and an information collection system. The Emergency Call Center receives various emergency telephone calls and cell phone

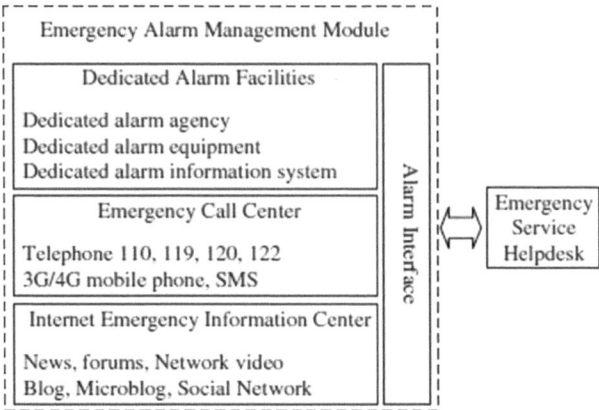

Fig. 2.10 Architecture of the emergency alarm management module

text messages. The Internet Emergency Information Center collects emergency information and proposals from the Internet. Alarm Interface handles the interactive information between this module and the Service Helpdesk Module.

The alarm management system can use RFID (Radio Frequency Identification), or other electronic tag information storage and marking functions with GPS satellite navigation and positioning technology, for monitoring and commanding storage and freight vehicles for emergency rescue supplies to form a "point-line-plane" real time weather information network. This system achieves precise positioning of freight vehicles, dynamic tracking, process control, and visualization management through the seamless integration of GIS and GPS location information. GPRS (General Packet Radio Service) communicates across the entire regulatory platform to achieve real time monitoring management of disaster rescue and disaster mitigation. This enables timely and reliable alarm and emergency calls when the GPRS identifies freight route deviation, freight vehicles pause timeout, and other abnormal conditions, thus ensuring safe and efficient delivery of rescue supplies.

2.11 The Problem Analysis and Management Module

Most emergency response activities cease when an emergency is over. However, there is no guarantee that the causes of the emergency have been identified and measures taken to prevent such emergencies in the future. In fact, similar emergencies may occur again. Therefore, emergency researchers must analyze the causes of emergencies and defects in the emergency response processes that were put in place. The Problem Analysis and Management Module in an ERDSS runs these analyses.

The Problem Analysis and Management Module provides the following services:

1. Accepts problem analysis instructions or requests;
2. Classifies and defines emergency problems;
3. Investigates, diagnoses, and analyzes problems;
4. Manages analyses and proposals;
5. Manages problem solutions and monitors implementation.

The Problem Analysis and Management Module corresponds to the Problem Management process in ITIL. The model's architecture is shown in Fig. 2.11.

The Problem Analysis and Management Module (see Fig. 2.11) has five functional components: Problem Classification and Control, Cause Analysis and Research, Solution Design and Evaluation, Report and Proposal, and Problem Interface. Problem Classification and Control identifies and diagnoses problems. Cause Analysis and Research analyzes and evaluates the causes of problems. Solution Design and Evaluation designs and evaluates problem solutions. Report

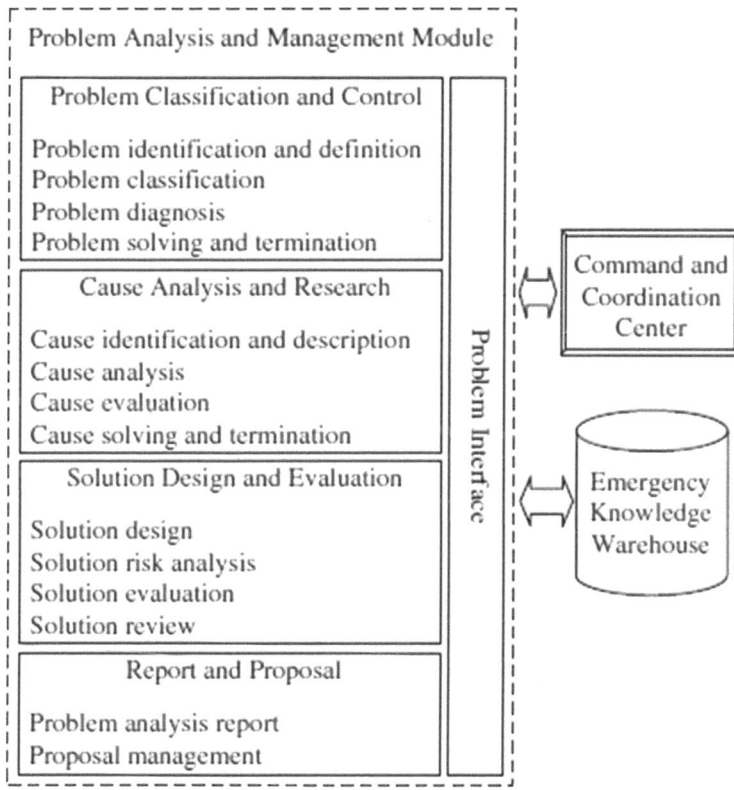

Fig. 2.11 Architecture of the problem analysis and management module

and Proposal generates formatted reports and proposals and submits them to the authorities. Problem Interface is a window that allows this module to interact with other modules.

The Q & A community studies natural disasters and investigates the characteristics of the problem and the answer. The theory and method of collecting characteristics relate to the quality of answers, in order to start from a topic and select the highest quality information. Text features and the characteristics of the indicators are selected and analyzed; the purpose is to evaluate the quality of the answers automatically. User satisfaction is the criterion for measuring the quality of the answers.

References

Alexander D (2005) Towards the development of a standard in emergency planning. Disaster Prev Manage 14(2):158–175

Barfod MB, Kim BS, Leleur S (2011) Composite decision support by combining cost-benefit and multi-criteria decision analysis. Decis Support Syst 51(1):167–175

Calixto E, Larouvere EL (2010) The regional emergency plan requirement: application of the best practices to the Brazilian case. Saf Sci 48(8):991–999

Cheng CY, Qian X (2010) Evaluation of emergency planning for water pollution incidents in reservoir based on fuzzy comprehensive assessment. Procedia Environ Sci 2:566–570

Green LV, Kolesar PJ (2004) Improving emergency responsiveness with management science. Manage Sci 50(8):1001–1014

Hernandez JZ, Serrano JM (2001) Knowledge-based models for emergency management systems. Expert Syst Appl 20(2):173–186

Ingram A (2010) Governmentality and security in the US president's emergency plan for AIDS relief (PEPFAR). Geoforum 41(4):607–616

Lu Y, Yang D (2011) Information exchange in virtual communities under extreme disaster conditions. Decis Support Syst 50(2):529–538

Lykourentzou I, Dagka F, Papadaki K, Lepouras G, Vassilaskis C (2012) Wikis in enterprise settings: a survey. Enterp Inf Syst 6(1):1–53

Schaafstal AM, Johnston JH, Oser RL (2001) Training teams for emergency management. Comput Human Behav 17(5–6):615–626

Siassakos D, Fox R, Crofts JF, Hunt LP, Winter C, Draycott TJ (2011) The management of a simulated emergency: better teamwork, better performance. Resuscitation 82(2):203–206

Chapter 3
ERDSS Key Technologies

Emergency management encompasses developing the necessary coping mechanisms and planning by government and other public institutions for pre-emergency warning, incident response, disposal, and post-incidence recovery. It makes use of technology and management tools to coordinate all available resources and takes a series of measures to protect public life, health, and property. An ERDSS uses computer, network communication, and space information technologies and other modern information technology to provide ancillary support for emergency management and decision-making, to improve the efficiency of emergency response and resolution. We have extracted the many techniques described below from the practical application of classical and effective models and algorithms. These technologies are effective in emergency management, particularly in providing new ideas and assisting in emergency response decision-making.

3.1 Disaster Assessment Technology

3.1.1 Assessment Indicator Framework

The proposed User-Generated Content (UGC)-based dynamic assessment framework for urban disaster indicators is based primarily on two aspects: assessing indicators of traditional disaster loss and their major and minor aspects and the characteristics of Internet UGC. In this context, the assessment indicator framework is divided into two parts: UGC direct loss indicators and UGC emotion converting loss indicators.

© The Author(s) 2017
S. Shan and Q. Yan, *Emergency Response Decision Support System*,
SpringerBriefs in Business, DOI 10.1007/978-981-10-3542-5_3

This chapter analyzes UGC content related to urban disaster, identifies loss assessment indicators according to the extraction algorithm, weights each indicator in the framework, and conducts a dynamic assessment of disaster losses overall.

3.1.2   Assessment Indicator Algorithm

We use the keyword matching method primarily for UGC direct loss indicators, set the appropriate keyword patterns, and weigh the number of matching results. In the period after the disaster, there is hysteresis (lag) in the description of the UGC content of direct loss indicators; therefore, the score is corrected by adding a time weighting to arrive at the final evaluation value of each item on the index. Keywords are selected if the content contains information such as "count + death" or "death + count" that matches communication patterns. The assessment of human loss indicators is calculated as follows:

$$score_p = \text{count\_p} \times \text{weight}_{time} \times \text{weight}_p \tag{3.1}$$

The assessment of financial loss indicators is calculated as follows:

$$score_t = \text{count\_t} \times \text{weight}_{time} \times \text{weight}_t \tag{3.2}$$

UGC event attention indicators are divided into two categories: efficiency indicators and number indicators of the UGC event. The increment is calculated as follows:

$$score = \sum_{i=0}^{n} \text{count}_{temp\_t[i]} \times \text{weight}_{ugc\_temp\_t}[i] \tag{3.3}$$

The number indicator of the UGC event largely describes the hard targets of disaster attention and is calculated as follows:

$$score = \text{count}_{all\_t} \times \text{weight}_{ugc\_all\_t} \tag{3.4}$$

UGC content related to disasters often contains a large amount of authors' emotional information, such as fear, compassion, trepidation, sadness, etc. Emotional information for individual disasters often carries related examples of losses from the disaster. These emotions are quantified and the main content of UGC emotion converting loss indicators extracted. The total emotion score is calculated as follows:

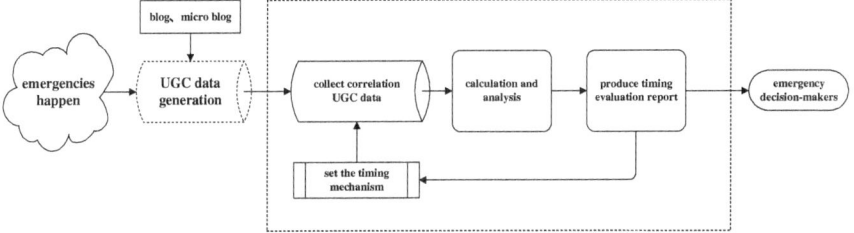

Fig. 3.1 Evaluation flow chart

$$score_{emotion\_total} = score_{emotion\_p} \times weitht_{emotion\_p} - score_{emotion\_n}$$
$$\times weight_{emotion\_n} \tag{3.5}$$

3.1.3   Evaluation Process

The process of UGC-based urban disaster dynamic assessment has two parts: data acquisition and data analysis. The evaluation process is shown in Fig. 3.1.

3.2   Adaptive Information Evaluation Technology

Blog texts all have a contextual index value for specific needs. There may be differences among the usage requirements for types of information for emergency response decision-makers. The demand unit is the type of emergency demand information required. The tendency of demand unit information for decision-makers is the weight of current work corresponding to the demand unit; the information demand of the current stage is obtained by adding up the weights of each information demand unit. The decision maker will use a large amount of blog information to assist in decision making. The actual use of information represents decision makers' current information demand tendency. Therefore, information demand distribution after decision makers actually decide serves as an indication of demand tendency between information units and work phases. The information requirements for each element are identified in several blog texts after mining for themes and screening for availability.

Specifically, blog information for decision-makers will be calculated on the contextual index value under each demand unit (D_i) of each blog to obtain the

Fig. 3.2 Adaptive evaluation
process employing demand
tendency

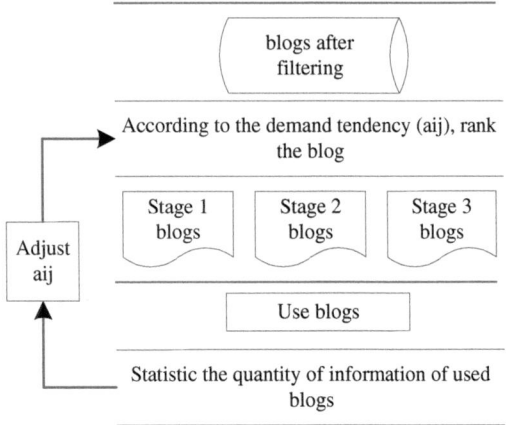

statistical distribution of the information and the proportion of total information used (a_{ij}) per demand unit (D_i) in this stage. (S_j) is the information demand tendency for a demand unit (D_i) in this stage (S_j). Thus, there will be differences in the possible use of blog texts.

$$P_{hj} = \sum_{i \in ISet_h} a_{ij} \times R_{hi} \tag{3.6}$$

where i represents the number of demand units; j represents the number of work stages; $ISet_h$ represents a collection of demand units containing the first h article or document; P_{hj} represents the degree of priority of the first h article blog in the j-th work stage; a_{ij} represents the information demand tendency to the i-th demand unit in the j-th work stage; and R_{hi} represents the degree of content relevance to the i-th demand unit of the first h article or document.

With respect to more than one decision, the blog information used in every decision updates the information demand tendency. Blog information is used in sequence according to the evolving demand tendency. This enables decision makers to use blog information quickly, conveniently, and efficiently, thereby creating an adaptive evaluation of the blog. The process is shown in Fig. 3.2.

In summary, the Blog Adaptive Assessment Model of demand for emergency response decision-making employs LDA (Latent Dirichlet Allocation) mining of themes to create the information demand units, applies GRNN (Generalized Regression Neural Network) to filter the blog according to usability evaluation indicators, generates blog priority according to information demand tendency, and updates information demand tendency according to the actual use of blog information in decisions. The process is shown in Fig. 3.3.

Fig. 3.3 Demand-oriented adaptive blog evaluation model

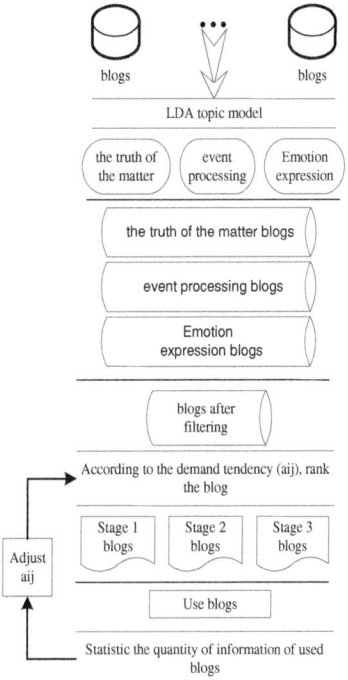

3.3 Knowledge Management Technology

3.3.1 Constructing Ontologies

Ontology modeling for the flood emergency database domain expresses knowledge related to the relationship between flood and flood emergency response using ontology in strict accordance with the principle of minimum body agreement. The ontology is constructed using the simple knowledge engineering method.

Before the establishment of the body, we must first determine the scope of the body in order to determine the concept, attributes, and relations that the body field contains. The database for the ontology domain of flood emergency is based on flood types and locations, disaster severity, affected areas, disaster intensity (number of old houses collapsed, number of people affected, mountain landslide situation, rainfall, and reservoir water level), causes of the disaster, and disaster duration. The strategic and tactical dimensions of emergency response for the flood emergency domain include rescue teams (transport sector, publicity departments, emergency command center, rescue department, other rescue teams) and rescue resources (emergency equipment, relief supplies, and materials for daily living).

From the perspective of ontology classification analysis, the flood emergency response body should include top-level, domain, task, and application ontologies simultaneously. The domain ontology of the flood emergency database should include geography, meteorology, rescue teams, materials and so on. We screened the content analysis of disaster emergency rescue, lessons from the above-mentioned study of earlier disasters, and terminologies in the field, in accordance with the classification system of concepts in the field and the composite structures of concepts, to obtain five flood emergency database fields: flood type, flood property, flood phenomenon, and disaster rescue.

Flood type data are qualitative—the flood belongs one of the categories of city flooding disasters, floods in mountain areas, reservoir and river floods, and sub-urban and town flooding disasters. Flood property data concern basic flood information, comprising affected areas, place, time, and disaster rating. Data about the flood phenomenon is particular information about individual floods, comprising disaster strength (damaged houses, number of people affected, water depth on the road, mountain landslide, rainfall, and reservoir water level), and the causes of the disaster and its duration. Disaster rescue data is emergency response after the floods, including rescue resources (emergency equipment, relief supplies, and materials for daily living) and rescue teams (transportation department, publicity department, meteorological department, communications department, etc.).

Overall, the flood relief database consists of two parts, namely the flood description section and the section describing the rescue program. Each includes a body portion and two conceptual levels that form the system for describing the flood relief database.

3.3.2   Case Similarity Algorithm

A combination of Euclidean distance and Cosine distance is used to calculate case similarity. Case similarity is measured from the two perspectives of absolute numbers and relative proportions.

3.3.2.1   Similarity Based on Calculated Euclidean Distance

The distance is calculated between two points in Euclidean space, assuming that X, Y are two points in n-dimensional space. The Euclidean distance between them is:

$$D(X, Y) = \sqrt[2]{\sum (x_i - y_i)^2} \tag{3.7}$$

As can be seen, when n = 2 the Euclidean distance is the distance between two points on a plane. The following formula is normally used for conversion when calculating similarity using Euclidean distance: the smaller the distance, the greater the similarity.

$$sim(X, Y) = \frac{1}{1 + d(X, Y)} \tag{3.8}$$

Principle: similarity s is defined by Euclidean distance d, s = 1/(1 + d).

Range: [0, 1]. The greater the value, the smaller d is. That is to say, the shorter the distance, the greater the similarity.

Description: As is the case with the Pearson correlation coefficient, this similarity does not consider the impact of overlapping numbers on the results.

3.3.2.2 Similarity Based on Calculated Cosine Distance

Cosine similarity is widely used to calculate the similarity of document data:

$$T(X, Y) = \frac{x \cdot y}{\|x\|^2 \times \|y\|^2} = \frac{\sum x_i y_i}{\sqrt[2]{\sum x_i^2} \sqrt[2]{\sum y_i^2}} \tag{3.9}$$

Principle: cosine value is calculated by the angle between two points in multi-dimensional space and the set point.

Range: [−1, 1]. The larger the value, the larger the angle. The farther apart the two points, the less the similarity.

Description: In mathematical expressions, if the properties of two items are at the data center, cosine similarity and the Pearson correlation coefficient are the same.

3.4 Knowledge Evaluation Technology

3.4.1 User Satisfaction and Artificial Quality Evaluation

The quality of information needs to be evaluated against certain criteria. Artificial assessment evaluation methods are divided into different categories according to the criteria they use. Evaluation criteria must be determined before the answers from community sites are targeted and evaluated.

On most community sites, questioners often have final discretion to determine the quality of the answers. Social and emotional content and effectiveness are the main considerations in questioners' choice of answers. If the answer does not meet the questioner's needs, the answers tend to be rated as lower quality and therefore the questioner's need is often the most important factor in the assessment of the answer. However, there are problems with these evaluation criteria, as community sites must meet the questioner's needs and the needs of other users. Moreover, the questioner's final choice is likely to be one-sided, highly subjective, and not the most effective answer. In addition, there are differences between the modes of evaluation used for different topic classifications. Users' choices, net friends' recommendations, expert advice, and other standards should be taken into account in multi-dimensional assessment.

3.4.2   Information Quality Classification and Evaluation Methods

3.4.2.1   Support Vector Machine (SVM) Algorithm

The basic concept of SVM is the two-dimensional case. When the case of a data set is linearly separable, the optimal classification surface is the surface of the largest interval.

Suppose that the sample set given is $\{(x_1, y_1), (x_2, y_2), \ldots (x_l, y_l)\}$, where $x_i \in R^d$, $y_i \in \{\pm 1\}$. After normalization, the equation of classification line H with H_1, H_2 is expressed as the following form.

$$H: w \cdot x + b = 0 \tag{3.10}$$

$$H_1: w \cdot x - b = 1 \tag{3.11}$$

$$H_2: w \cdot x - b = -1 \tag{3.12}$$

Seeking the optimal classification line to solve the problem is expressed as a quadratic programming problem:

$$\text{Min } f(\omega, b) = \frac{1}{2} ||\omega||^2 \tag{3.13}$$

$$\text{s.t. } y_i(\omega \cdot x_i + b) - 1 \geq 0 \quad i = 1, 2, \ldots, l \tag{3.14}$$

Optimal classification functions after solving the above problems are as follows:

$$f(x) = \text{sgn}\{\omega \cdot x + b\} = \text{sgn}\{\sum_{i=1}^{m} \alpha_i^* y_i(x_i \cdot x) + b^*\} \tag{3.15}$$

In the above formula, m represents the sum including only the support vectors, i.e., the corresponding sample vectors when the Lagrange multipliers are not zero. In addition, b* is the classification threshold, which can be obtained with any one of the support vectors, or by the median value of any pair of support vectors calculable in two categories.

The optimal classification surface can be calculated only on the premise of linear separability. Some training samples cannot satisfy the constraint of linear separability; in those cases, the slack variables ξ_i and penalty factor C need to be introduced in order to calculate the optimal generalized classification surface without requiring a zero training error rate, but maximizing the samples. This will ensure the maximum partition interval. Introducing slack variables and the penalty factor, the quadratic programming problem is shown as follows:

$$\text{Min } f(\omega, b) = \frac{1}{2}||\omega||^2 + C\sum_{i=1}^{l} \xi_i \tag{3.16}$$

$$\text{s.t. } y_i(\omega \cdot x_i + b) - 1 + \xi_i \geq 0 \quad i = 1, 2, \ldots, l \tag{3.17}$$

In the above formulas, C is a constant, adjusting the punishment degree for separating samples incorrectly and achieving algorithm reconciliation between tolerance of incorrect sample separation and computational complexity.

3.4.2.2 Logistic Regression

The logistic regression model is a classification algorithm with values 0 and 1 that calculates the probability of a particular item being attributable to two categories based on the input of a set of independent variables. The logistic regression model commonly uses binomial logistic regression. The conditional probability distribution in binomial logistic regression is as follows:

$$P(Y = 1|x, \omega) = \frac{e^{\omega^T x + b}}{1 + e^{\omega^T x + b}} = \frac{1}{1 + e^{-(\omega^T x + b)}} \tag{3.18}$$

$$P(Y = 0|x, \omega) = \frac{1}{1 + e^{\omega^T x + b}} \tag{3.19}$$

where $x \in R^n$ is the input and $Y \in \{0, 1\}$ is the output. The two classifications may correspond to "high-quality answer" and "low quality answer." Function parameters are $\omega \in R^n$ and $b \in R$; ω is the vector of weights. Each value of the vectors corresponds to the weight of each input feature. B is the bias.

3.5 Blog Evaluation Technology

3.5.1 Blog Theme Recognition in the Emergency Response Category

The main purpose of theme recognition is to determine whether the blog is concerned about emergency response by identifying the blog's theme. We need to capture data from blog sites, analyze the website to identify if the content is relevant, and preprocess the data. First, blog data is captured and collected. The second step uses the Vector Space Model (VSM) to conduct Chinese Word Segmentation on the blog content and expresses blog content in word vectors so the text can be processed as a structured form. The third step conducts cluster analysis of the blog text. The blog set is clustered into one blog cluster to get the central value ("centroid") of the blog cluster, and selecting several high-weight words are selected as the keywords of the blog cluster. Finally, the keywords are compared with the thesaurus of our emergency response. If the similarity between the two is high, the blog cluster belongs to the theme of emergency response and all blogs in the blog cluster are considered to be in line with the theme of emergency response.

3.5.2   Blog Evaluation Model Based on Information Entropy

This approach builds a framework based on information entropy to evaluate the value of emergency response blogs for emergencies. Suppose U is a set of evaluation indicators representing the blog information. U_1, U_2, ..., U_n represents every aspect of the evaluation and each evaluation indicator. Then $U = U_1 + U_2 + \cdots + U_n = \sum U_i$ (i = 1, 2, ..., n). The probability of each indicator is $P_i = U_i/U = U_i$, and $\sum P_i = 1$, where P_i represents the possibility of every evaluated aspect. Therefore, the information entropy of each blog is defined as:

$$H = -\sum P_i \log P_i \qquad (3.20)$$

where H represents the entropy of each piece of information. The larger the differences between the possibility of every aspect P_i is for each piece of information, the greater the entropy H of this information, representing the greater "value" of this information. When the possibility of every aspect equals P_i, information entropy reaches the theoretical maximum. The maximum entropy is fixed for any given information evaluation system.

In this section, the value of the blog is analyzed primarily from the perspectives of correlation, consistency, semantics, and sociality (three dimensions).

When evaluating emergency response blogs about an incident, evaluation values are normalized to ensure that they are dimensionless (after obtaining the evaluation values in accordance with the indicators in the framework above), and substituted into the formula for calculation. Information entropy is obtained as a comprehensive evaluation value and used as the basis for blog evaluation.

3.6   Information Dissemination Analysis Technology

3.6.1   Emergency Information Evaluation in the Virtual Community

Emergency information in the virtual community has particular characteristics that frequently affect its propagation. Therefore, this chapter proposes an evaluation framework that takes propagation factors into account and applies them to emergency information after analyzing the features of emergency information in the virtual community. Information features are described quantitatively below to provide a basis for an examination of the operating rules for propagation (propagation discipline).

3.6.2 Evaluation Framework for Emergency Information Propagation

To study propagation discipline that takes account of the different characteristics of emergency information in the virtual community, we build a framework based on information entropy to evaluate the value of information propagation. Suppose that U is a collection of indicators about emergency information in the virtual community, U_1, U_2, ..., U_n represents every aspect of the evaluation—in other words, every evaluation indicator. Then $U = U_1 + U_2 + \cdots + U_n = \sum U_i$ (i = 1, 2, ..., n). The probability of each indicator is $P_i = U_i/U = U_i$, and $\sum P_i = 1$, where P_i represents the possibility that every aspect has been evaluated. Therefore, the information entropy of each piece of emergency information in the virtual community is defined as:

$$H = -\sum P_i \log P_i \qquad (3.21)$$

where H represents the entropy of each piece of information. For each piece of information, the larger the differences between the possibilities of every aspect of P_i, the greater the entropy H of this information, representing its greater "value." When the possibility of every aspect P_i are equal, information entropy reaches the theoretical maximum. The maximum entropy is fixed for any given information evaluation system.

3.6.3 Emergency Propagation Model for the Virtual Community

The innovation diffusion model is a series of models that estimate the number of innovation products evolved over time. It is also applicable to research into virtual products, policy innovation, and other areas. Therefore, innovation diffusion models are useful for building a suitable model for emergency information in the virtual community. In Sect. 3.6.3, we proposed that the value of emergency information dissemination in the virtual community is based on information entropy theory. In this section, we will extend our analysis of the propagation discipline of emergency information in the virtual community on this basis. First, we classify emergency information in the virtual community according to the scale of information entropy. Next, we use the innovation diffusion model to model the time diffusion and propagation discipline of emergency information in the virtual community of the results.

Emergency information propagation is closely linked to emergency types, time of occurrence, and the information itself. Therefore, we consider diffusion discipline as well as the type of emergency and time of occurrence separately for each different type of information. We use the innovation diffusion model to compare differences between the diffusion disciplines of different types of information.

The diffusion of emergency information in the virtual community benefits from hot spots in the information, the author's influence in the virtual community, and his/her contribution to emergency warning and rescue efforts. However, the metrics of these factors and information entropy are largely consistent. Therefore, it is possible to establish a logistic model, Gompertz model, or Bass model to study the diffusion discipline of emergency information based on the information's entropy classification.

3.7 Emergency Resource Scheduling Technology

3.7.1 Modeling the Urban Emergency Supplies Scheduling Process

The "emergency" aspect of the emergency supplies scheduling process occurs when cities abruptly experience a natural disaster. We employ the same structure as the earlier part of this paper to describe the functions and logical relationships in the supplies scheduling process. We then construct a city emergency supplies scheduling petri net general model, employ a collaborative detection algorithm to identify whether there is conflict, and build a performance analysis algorithm based on the Markov chain.

3.7.2 Unit Analysis of the Emergency Resource Scheduling Process

We employ strategic research into urban emergency resources scheduling and critical process analysis to divide the complete emergency resource scheduling process roughly into preparation, implementation, evaluation—three stages altogether. We summarize all the functions in the emergency resource scheduling process according to participants' different roles, responsibilities, and tasks. The preparation stage involves a patching unit, a disaster real-time monitoring unit, a resource demand forecasting unit, and an emergency resource mobilization unit.

The implementation stage involves a control unit, an identification unit, a transportation unit, a deployment unit, and a feedback unit. The evaluation stage involves an investigation unit, an assessment summary unit, and an information unit that operate across all the stages.

3.7.3 Establishing the Petri Net Model

Petri net collaboration is primarily concerned with order, parallel, and choice collaboration. We use the three collaboration methods to combine the units summarized in the previous section according to their orders and logic relationships to form the Petri net city emergency supplies scheduling model shown in Fig. 3.4.

As shown in Fig. 3.4, the starting point of this model is the time at which emergency disaster information is received. The first step is to establish a temporary dispatch center; the disaster information acquisition unit begins operations immediately, gathering information, broadly divided into meteorology, ground, and water information, simultaneously; after monitoring of all three areas is complete, the information acquisition link ends and the demand forecasting link begins. The demand forecasting unit provides the necessary scientifically based reference data for supplies financing. Once this link ends, the emergency resource financing unit can carry out its tasks. The unit consists of three parallel sub-tasks: resource scheduling, donor collection, and organizing supply production. After the emergency resource mobilization program is established, the control body begins dispatching personnel and distributing material. Therefore, the demand prediction unit, the emergency resource mobilization unit, and the control unit are in a synergistic relationship. During deployment, scheduling from supply point to the transit center and from the transit center to the affected point should be an order relationship, and the feedback unit monitors the implementation of the whole supply scheduling process. The protection unit also controls the logistics content of the entire deployment process, so when the unit begins deploying supplies, the monitoring and protection units start up at the same time. Troubleshooting commences once

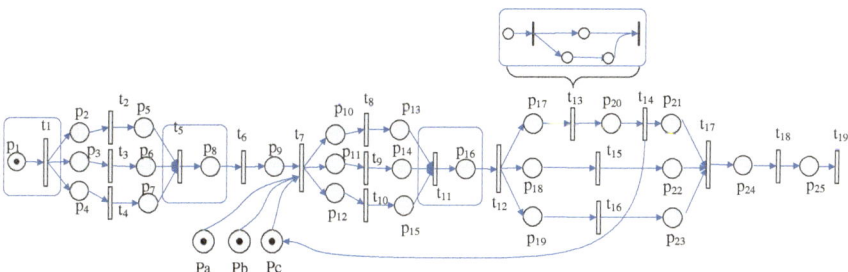

Fig. 3.4 Urban emergency supplies scheduling petri net model

deployment is complete. Troubleshooting feedback is delivered to the command center for supply allocation for the secondary deployment plan, so the arrow here points to the supply financing library. When investigation is complete, the feedback and protection units also complete their work. Once the main task of the investigation is finished and the investigation and report of the secondary material scheduling needs are completed, we need to summarize and assess the process.

3.8   Simulation Modeling Technology

3.8.1   Analysis of Agent Structures in the Emergency Plan

As urban natural disaster emergency plan involve many responsible organizations operating in an environment of dynamic uncertainty across multi-level structures. A single task may involve more than one emergency organization, and one emergency organization may need to undertake multiple tasks. The network association structure between tasks and emergency organizations determines the complexity of our model.

We use the concept of an agent to simulate the actions of organizations involved in emergency plans, the internal structure of the agent to describe their responsibilities and participation in the process of emergency response, and hybrid architecture agents to build the general agent structural model for the urban natural disaster emergency plan. An emergency response agent generally consists of five functions: environmental perception, analytical reasoning, decision making, undertaking tasks, and collaborative interaction.

The main environmental perception task is to collect and process information from outside the organization; analytical reasoning uses primarily model, method, and knowledge bases to infer solutions to problems. Analytical reasoning is used for information analysis, evolutionary reasoning, calculation, modeling, and so on. Decision-making's primary function is to achieve the agent's mission objectives using the results of the analysis. This requires task decomposition, goal planning, and strategy matching, and so on. Undertaking the task puts the decision plan into practice in order to complete the mission objectives, and collaborative interaction's primary role is to maintain the interactive communication between the agent and other agents for information transmission.

3.8.2   Petri Net-Based Emergency Plan Process Modeling

The Petri net model of emergency response decision making meets the following basic requirements: comprehensibility, correctness, applicability, static and dynamic *utility*, and enforceability of the model. It ensures the integrity and internal relations of the system based on an accurate description of business processes. Petri net is used to model the relationships among all the emergency organization agents involved in the process model, taking account of the characteristics and relationships among various sections' processes. The central core is analysis, decision-making, directing, execution, control, aftermath disposal, evaluation, and summary before, during, and after the unexpected event—three stages altogether, that display the relationships among the various activities accurately.

The monitoring agent starts its tasks when the emergency plan process begins. The monitoring agent is divided into meteorological, ground, and water monitoring agents, with parallel task execution; in other words, collaborative relationships begin with parallel collaboration. The analysis agent can begin analysis only when all three monitoring tasks are complete at the end of the monitoring mission; in other words, in this case the collaborative relationship ends parallel collaboration. The analysis agent provides the necessary basic data for the decision agent; after the analysis, the decision agent can conduct the decision tasks. After the program is established, the command agent begins the command tasks, analysis Agent, decision Agent. The command agent is order collaborative relationship; during the mission, the execution agent is divided into defense, rescue, supplies, health, and security agents, of which the rescue agent acts only after the defense agent ceases. When mass transfer is complete, we can expand the rescue work, while other supplies, health, and security agents can begin in parallel.

When all tasks are finished, it is the end; that is to say, the end of the transition is the end of parallel collaboration; the supervisory agent supervises the implementation of the mission, so when the task agent begins, the supervisory agent begins; when the mission is completed the supervisory agent ends. The supervisory agent and the task execution agent execute their tasks in parallel collaboration. Similarly, in the aftermath of task execution, reconstruction and rescue mission occur only when the investigation agent's task is finished and investigation into all hidden dangers from the disaster is completed. The reconstruction agent and the rescue agent begin parallel collaboration; when the rehabilitation task is completed, it is time to summarize and assess the process.

3.9 Other Related Technologies

Unexpected natural disasters and accidents are always critical issues for people to respond to; how to prevent and deal with these issues effectively is the core of the work of emergency management. Damage assessment throughout the emergency management process also plays a very important role. Pre-assessment before the disaster, dynamic assessment at the time of disaster, and overall assessment of the disaster, are all critical to decision-making in emergency management. The following techniques and algorithms summarize some of the other approaches proposed by other scholars that may be used in emergency response processes.

3.9.1 Dijkstra's Algorithm

Shendarkar et al. (2008) have introduced the shortest path algorithm in their work, which presents a novel methodology involving a Virtual Reality (VR)-based Belief, Desire, and Intention (BDI) software agent to construct crowd simulation and to demonstrate its use in crowd evacuation management during terrorist bomb attacks in public areas. They employ Dijkstra's shortest path algorithm to calculate the shortest path. They have selected Dijkstra's algorithm because of its simplicity and quadratic computational cost.

1. Definition

Dijkstra's algorithm (2010), conceived by Dutch computer scientist Edsger Dijkstra in 1956 and published in 1959, is a graph search algorithm that solves the single-source shortest path problem for a graph with non-negative edge path costs, producing a shortest path tree. This algorithm is used frequently in routing and as a subroutine in other graph algorithms.

For a given source vertex (node) in the graph, the algorithm finds the path with lowest cost (i.e., the shortest path) between that vertex and every other vertex. It also finds the cost of the shortest path from a single vertex to a single destination vertex by stopping the algorithm once the shortest path to the destination vertex has been determined. For example, if the vertices of the graph represent cities and edge path costs represent driving distances between pairs of cities connected by a direct road, Dijkstra's algorithm finds the shortest route between one city and all other cities.

2. Algorithm

Let the starting node be called the initial node. Let the distance of node Y be the distance from the initial node to Y. Dijkstra's algorithm will assign some initial distance values and improve them systematically.

(1) Assign to every node a tentative distance value: set it to zero for our initial node and to infinity for all other nodes.
(2) Mark all nodes unvisited. Set the initial node as current. Create a set of the unvisited nodes called the unvisited set consisting of all the nodes except the initial node.
(3) For the current node, consider all of its unvisited neighbors and calculate their tentative distances. For example, if the current node A is marked with a tentative distance of 6, and the edge connecting it with a neighbor B has length 2, then the distance to B (through A) will be $6 + 2 = 8$. If this distance is less than the previously recorded tentative distance of B, then overwrite that distance. Even though a neighbor has been examined, it is not marked as visited at this time, and it remains in the unvisited set.

(4) When we are done considering all of the neighbors of the current node, mark the current node as visited and remove it from the unvisited set. A visited node will never be checked again; its distance recorded now is final and minimal.
(5) If the destination node has been marked visited (when planning a route between two specific nodes) or if the smallest tentative distance among the nodes in the unvisited set is infinity (when planning a complete traversal), then stop. The algorithm has finished.
(6) Set the unvisited node marked with the smallest tentative distance as the next "current node" and go back to step three.

3.9.2 Ant Colony Optimization Algorithm

Yi et al. (Sheu 2007) proposed an ant colony optimization (ACO)-based heuristic algorithm for the multi-commodity and vehicle dispatching problems that arise in disaster relief activities. The distinctive feature of their method is that the original emergency logistics problem is decomposed into two phases of decision making: vehicle routes construction and the multi-commodity dispatch, where vehicles, relief, and wounded people are treated as commodities, formulated as a multi-commodity network flow problem in the second phase, and solved using the proposed ACO meta-heuristic algorithm.

The ACO algorithm is a probabilistic technique for solving computational problems that can be reduced to finding good paths through graphs in computer science and operations research.

In the natural world, ants wander randomly, and upon finding food return to their colony while laying down pheromone trails. If other ants find such a path, they are likely not to keep travelling at random, but instead follow the trail, returning and reinforcing it if they eventually find food.

Over time, however, the pheromone trail begins to evaporate, reducing its attractive strength. The more time it takes an ant to travel down the path and back, the more time the pheromones have to evaporate. A short path, by comparison, is marched over more frequently, and thus pheromone density becomes greater on shorter paths than longer ones. Pheromone evaporation also has the advantage of avoiding convergence to a locally optimal solution. If there were no evaporation, paths chosen by the first ants would tend to be excessively attractive to the following ones, and exploration of the solution space would be constrained.

Thus, when one ant finds a good path from the colony to a food source, other ants are more likely to follow that path, and positive feedback eventually leads all the ants to follow a single path. The idea of the ant colony algorithm is to mimic this behavior with "simulated ants" walking around the graph representing the problem to solve.

The original idea comes from observing the exploitation of food resources among ants, in which ants' individually limited cognitive abilities are able collectively to find the shortest path between a food source and the nest.

1. The first ant finds the food source (F), via any way (a), then returns to the nest (N), leaving behind a trail pheromone (b).
2. Ants indiscriminately follow four possible ways, but the strengthening of the original way makes it more attractive as the shortest route.
3. Ants take the shortest route and long portions of other ways lose their trail pheromones.

Biologists have observed that ants tended to use the shortest route in a series of experiments on a colony of ants with a choice between two unequal-length paths leading to a food source (Deneubourg et al. 1990).

3.9.3   Network Terminal Reliability

The probability that at least one path exists between an origin and a destination after a disaster is defined as Network Terminal Reliability (NTR) or Connectivity Reliability (Victor and Boris 1996). Much research (Taniguchi et al. 2012) has been conducted to define indices and their application to NTR in the context of transportation; for instance, Nicholson (2007) has studied the application of NTR in identifying strategies to optimize resource allocation to maximize network reliability. NTR has also been proposed as a model for hazard loss estimation under theoretical paradigms of network reliability (Nicholson and Giovinazzi 2010). Lai (2011) suggested that NRT is useful in both "business-as-usual" and emergency situations. However, he also noted that unforeseeable/uncontrolled events (e.g., natural and man-made disasters) are of special concern as link failures can be estimated only but not defined precisely, as is usually the case when maintenance and construction works are scheduled.

NTR is potentially useful for Emergency Management. A good understanding of possible failures and impacts on individual links due to hazards is vital for emergency response and recovery. For a given road network, the overall network reliability (R) depends on the probability of failure of individual links.

Considering that the reliability (probability of success) for individual links can be defined by their respective expectation of survival after an event (i.e. $r = E[x]$), then the true value of NTR can be estimated. Since a road network is usually a complex structure, it is necessary generally to break it down into basic components, namely series, parallel, and bridge components.

The scientific literature includes further indicators and work related to NTR. Wakabayashi (2004) has proposed a Criticality Index and compared the results against Birnbaum's Reliability Importance Index, while Nicholson (2007) described and used Birnbaum's Index and the Criticality Index of Henley and Kumomoto

(1992). Both authors sought to identify which index or combination provided better support for decision-making to maximize network reliability by allocating resources to improve the reliability of individual links. Recently, Lai (2011) used all three indices to study a given road network in New Zealand to explore further the advantages and disadvantages identified by Nicholson and Wakabayashi in applying such indices to road networks.

3.9.4 Path Selection and Network Indices

Network connectivity is related inevitably to the number of existing paths between different origin and destination pairs and their individual relationships with the network as a whole. The number of possible paths is a function of the number of links and the network configuration, which depends upon the number of nodes.

Researchers have investigated a broad range of approaches and proposed a series of theories to describe network configuration. For instance, Akgün et al. (2000) studied path dissimilarity, Sharafat and Ma'rouzi (2009) proposed a Recursive Truncation Algorithm (this involves scanning minimal cut-sets to determine weak and strong cut-sets, and comparing the failure probabilities of weak cut-sets, while the failure probabilities of strong cut-sets are ignored), and Dijkstra (1959) long ago proposed an efficient algorithm for identifying the shortest path.

Network indices that quantify and compare elements of a network are also found in the scientific literature. Lai (2011) presents a brief discussion of some indices, such as road network performance (Scott et al. 2006), the network density index (DeLaurentis et al. 2008), the network connectivity index [which considers the relationship between the number of links and maximum number of paths (Scott et al. 2006)], and the network robustness index (Scott et al. 2006), which evaluates the critical importance of a given highway segment to the overall system as the change in travel time cost associated with re-routing traffic.

We selected a combination of two indices from the above: a variation of Dijkstra's shortest path algorithm, which identifies the shortest paths and the network connectivity index proposed by Scott et al. (2006). The latter involves calculating the computed gamma index (γ). Note that gamma can range between zero and 1, where 1 means a completely connected network and emax is also calculated.

3.9.5 Cost Estimation

Costs are usually employed in transportation research to compare the efficiency of different planning or management decisions. Ferreira (2010) proposed two cost functions in the particular context of emergency management to optimize resource allocation, namely Logistics Response Costs (LRC) and Delay Response Costs (DRC). Logistics costs are associated with resource mobilization from a given origin to a destination, while delay costs are an estimate of traffic delay due to total or partial road closure because of emergencies.

3.9.6   Artificial Neural Networks (ANNs)

Artificial Neural Networks are computer-based algorithms inspired by the structure and behavior of real neurons.

An ANN consists of a set of processing units that simulate neurons interconnected via a set of links allowing signals to travel in parallel and serially. Each link multiplies the output from a unit by a weighting factor, a value analogous to the connection strength at a synapse. The link then passes the weighted output value to another unit, which sums up the values passed to it by all other incoming links. If the total input value exceeds a designated threshold value, the unit fires. Modifications in firing patterns constitute learning.

An ANN is typically composed of three layers: input units, representing the information that is fed into the network; output units, representing the network output; and one or more hidden layers of units between them. In order to train an ANN to perform a certain task, we must first determine how the units are connected (the network architecture). Training involves presenting the neural network with specific examples (data sets). These consist of a pattern of input activities for the input units and the desired pattern of activities for the output units. The data sets are presented repeatedly to the network, and the weights are updated after each presentation to constitute learning. One of the most successful supervised training methods is the back-propagation algorithm. The basic concept is to use the derivative of an error function in order to find the direction that minimizes network error and update the weights accordingly.

Regarding the suitability of ANN systems for emergency response, Liao et al. (2012) proposed a general methodology for developing environmental emergency decision support systems based on an artificial neural network. The procedures for matching environmental emergency decision support characteristics were as follows. First, case information and emergency measures are digitized (coded): case information is divided into input attributes and decision-making information, and standardized and digitized using the Feature Evaluation (FE) method and the Intensity Hierarchical (IH) method, respectively. Second, an environmental emergency ANN is constructed, in which the Gradient Descent with Momentum and Adaptive Learning Rate (GDMALR) method, a modified back-propagation algorithm, is employed for training. Third, decision-making information is translated (decoded); ANN output data is interpreted into practical contingency measures using the Translation Based on Conventional Import Ratios (TBCIR) method. The author concluded that environmental emergency decision support systems based on ANN technology are effective and generate emergency plans quickly and accurately to meet the requirements of an environmental emergency on-site.

3.9.7   GIS (Geographic Information System)

A GIS (Geographic Information System) is a system that offers the functionality and tools to collect, store, retrieve, analyze, and display geographical-related information. In general, GIS can handle two types of data: vector and raster data. Vector data are defined by pairs of coordinates, and present extremely accurate coordinate geometric information with small data storage requirements. Features, events, and activities with a spatial component are modeled as points, lines, polygons, nets, or links to form the geographical relational database. Raster data are stored using a homogeneous grid system.

It is possible to design an interactive system using GIS technology that facilitates comprehensive disaster management. Ni-Bin Chang developed a disaster support system (DSS) configured to enhance the performance of decision makers responsible for chemical emergency preparedness and response. It demonstrated how GIS works with database, multi-scale model base, and risk assessment knowledge to create an interactive chemical emergency preparedness and response system (CEPRS). The capabilities of such a DSS are (a) analysis and visualization of inventory characteristics in chemical industrial complexes, (b) simulation and impact evaluation of release events, and (c) generation of response strategies based on risk assessment.

GIS maps have great potential in the management and planning of responses to crises involving the release of hazardous materials. Satellite images and remote sensing technology, with high resolution images and the latest sensors, are capable of providing a wide variety of information such as a synoptic view of the release location and surrounding land use, vulnerable locations susceptible to damage, terrain conditions, 3D urban structures, and possible escape routes.

The type of information that a GIS-based map offers is crucial for emergency response planning to make important information available in a simple, easy to understand manner for decision makers and officers responsible for crisis management, to enable them to understand fully the crisis's progress.

Alhajraf et al. (2005) proposed a GIS-based system that can assist an emergency response manager to plan effectively for the most suitable action for the case in hand. The system is useful at three distinct times: before an incident for better planning and preparedness, during an emergency for online real time follow-up of the situation on the ground, and after the incident for aftermath investigation. The system consists of the three main steps. The first step is gathering basic input requirements to run the system. The three major input data are:

1. Meteorological data close to the emission source. This data can be gathered from online weather stations or predicted from sophisticated weather forecasting models.
2. Information about the source including geographical location, type of material released and emission rate, height, size, and temperature.
3. Other supporting information such as topographical information, GIS map, and information about surrounding land use near the emission area.

In the second step, output results are prepared in a graphical, easy to understand form and superimposed over available satellite images or a GIS map.

The last step in the system is to forward the results to the right person:

1. Emergency response team at the department of civil defense;
2. Firefighting headquarters;
3. Hospitals and health service providers; and
4. Other relevant organizations.

3.9.8 Mobile Technology

After the earthquake in Haiti in January 2010, about a third of Haitians had access to mobile phones compared to 11% with access to the internet. Messages with personal live reports and geographic information transmitted through SMS, MMS, and the internet provided valuable information for helping rescue victims. Mobile devices' wide coverage offered many more chances for effective relief measures.

3.9.9 Immunology

Immunology is applied primarily in biology, medicine, artificial intelligence, computer network security, and mathematics. The human immune system has been modified considerably during the long process of evolution and is an effective and adaptive early warning and defense system. Purposeful and planned vaccination prompts the immune system to create a rapid and efficient immune response to specific antigens.

There are similarities and differences between the immune system and emergency management systems, but there is the probability of establishing an emergency management artificial immune system and of constructing active defense "bionic" systems that specify an emergency risk identifier and document the immunization mechanism in the emergency plan.

Qing et al. (2012) proposes such an approach. A multi-agent system (MAS) composed of multiple interacting intelligent agents within an environment is able to solve problems that are difficult or impossible for an individual agent or a monolithic system to solve. Search and analysis functions may include methodical, functional, procedural, or algorithmic search, find, and processing.

The antigen recognition mechanism implements the early-warning function through T cell and B cell identification of specific antigens, extraction of key determinants, delivering key information, and even identifying the key factor, reinforcing the recognition competence of the risk of sudden events through gene pool construction, and the evolution of identification devices.

The author concluded that antigen recognition in the traditional artificial immune system is completed by the identifier's adaptive and self-learning evolution. In active immune system engineering of emergency management, the risk identification process uses the identifier clonal selection model and optimization methods creatively, choosing among more than one recognition agent the one that has the greatest antigen affinity and eliminating the others. Experimental results show that through the integration of the immune and genetic algorithms, the evolutionary identifier can make risk (antigen) recognition more effective, accurate, and fast. Multi-agent immunological active defense emergency management systems with integrated algorithms, continuous self-learning, and evolution guarantee effective and timely active immunization. The premise of the normal operation of the system is to obtain the full raw data set and set the threshold value range.

3.9.10   Social Media

Managing knowledge is paramount for organizational survival and effectiveness in turbulent, fast changing environments. Organizations responding to disasters that affect them directly and indirectly are compelled to operate in such environments. However, disaster response often presents a unique situation that traditional knowledge management systems (KMS) are not optimally configured to support. Governments, aid organizations, businesses, and individuals alike face unique KM challenges when responding to a disaster.

"Communication and information systems in affected areas may be degraded or unavailable" (Yates and Paquette 2010 31(1):6–13). Tasks may be out of the ordinary or emergent. Resource constraints may require new ways to think about existing responsibilities and functions. Knowledge availability varies more extremely than in normal situations—sometimes little information will be available with which to make informed decisions; at other times, multiple reports with conflicting information may necessitate increased information processing capabilities. Finally, decision-making must occur in a compressed timeline since faster than usual response is needed to stabilize a dangerous situation, prevent further losses, and begin reconstruction. Knowledge management systems that will be useful in a disaster response must be flexible enough to handle unexpected situations yet robust enough to be reliable in degraded or complex environments.

Social media is one emerging technology with the potential to allow for the flexibility, adaptability, and boundary-spanning functionality demanded by response organizations for their information systems. Social media technologies have the ability to coordinate widespread communication and strengthen information flows and to be flexible to the changing needs of the responders.

Boyd and Ellison (2007) define social media as "web-based services that allow individuals to (1) construct a public or semi-public profile within a bounded system, (2) articulate a list of other users with whom they share a connection, and (3) view and traverse their list of connections and those made by others within the system."

The difference between these technologies and other standard forms of ICTs is that users are able to make their views, perceptions, and knowledge public via the system. This forms ties with other individuals who may have similar interests, needs, or problems. While networking may not be the primary motivator for their use, these technologies allow knowledge sharing through the creation of knowledge networks.

Social media can be, and often are, used to help consumers make informed product or service decisions; this is a relatively new business model that has not been adequately explored. Social media also provide a basis for organizational decision making using idea generation. Notably, social networking applications are beginning to be seen in critical decision-making fields. For example, social networking has been discussed as a tool to support decision making by pilots (Scott 2011) and decision making in healthcare (Griffin and de Leastar 2009). One of the current authors has a graduate student working on a system to integrate live feeds from Twitter and Facebook to provide current information to emergency response teams. These systems use a combination of the algorithms and structured data that traditionally make up a DSS, but add the evolving technology of GISs and social media to the models and information to aid decision making.

The Weather Channel captures social media feeds online so that people can see quickly what weather trends are across the country. Recent research has explored the relationships among gender, age, and a social networking site member's propensity to be affected by advertising through that venue (Taylor et al. 2011). These examples demonstrate that social media is a pervasive technology accepted by organizations as a viable platform; future research will explore its potential and limitations.

Trust is one continuing theme throughout DSS research that has particular relevance to social media. Researchers have long investigated the relationship between trust and information systems. The development of trust measures for e-commerce by McKnight et al. (2002) is important for social media. However, recent research indicates that what we thought we knew about trust in information systems artifacts may not hold true for social media. People tend to attach social characteristics to IS artifacts (Al-Natour et al. 2011). Because many social media users communicate primarily with individuals through a network of people whom they know personally, the idea of trusting social media itself may not be as important to the member as is trust in the member's friends.

Virtual worlds/communities are another relevant area of social media. Individual use Second Life and other communities as pastimes; businesses use them as test beds and training tools. The military has considered Second Life for training and recruiting (Cacas 2010).

Social media facilitate organizational knowledge sharing in two ways: by increasing knowledge reuse among employees and by eliminating the reliance on formal liaison structures (both personnel and systems) between employees. Prior to social media, most knowledge was shared during formal briefings (where little conversation was possible and there was little insight into how knowledge in each function was acquired and built); today, each employee has complete visibility into how their colleagues are managing knowledge. This includes access to their sources, identifying when different functions were working on the same problem from different ends (something that occurs frequently), and finding materials that could easily be repurposed for other needs. The other major advantage offered by using social media in the response was in the advantages gained from bypassing or eliminating formal liaison structures used previously to share knowledge between different agencies. Besides translating knowledge from one domain to another, the liaison's most valuable function was brokering knowledge sharing requests (Wenger 1998). Typically, the problem was not so much that staff in one agency were not allowed to access knowledge from another, but that it would be practically difficult for them to know who, what, where, and how to access that knowledge. Social media facilitated this awareness.

3.9.11 Information and Communications Technology (ICT)

Information and communications technology (ICT) is used frequently as an extended synonym for information technology (IT), but is a more specific term that stresses the role of unified communications and the integration of telecommunications (telephone lines and wireless signals) and computers, as well as necessary enterprise software, middleware, storage, and audio-visual systems. This enables users to access, store, transmit, and manipulate information.

The term ICT also refers to the convergence of audio-visual and telephone networks with computer networks through a single cabling or link system. There are large economic incentives (large cost savings due to elimination of the telephone network) to merge the audio-visual, building management, and telephone network with the computer network system, using a single unified system of cabling, signal distribution, and management.

ICTs are bringing about remarkable changes in response to emergencies. This includes areas of emergency medical service, structural and wild-land firefighting, urban search and rescue, emergency evacuation simulation, emergency dispatch work, and information systems modeling for formal response. High involvement by members of the public in emergencies is critical, and ICT makes their role more visible and broadens the scope of their participation. ICT-supported communications in emergency will result in at least three changing conditions that need to be addressed by the formal response. ICT-supported citizen communications can spawn, often opportunistically, information useful to the formal response effort.

The availability of these data means that people can seek and access further information from each other. Such capability can help the formal response effort in collecting and providing information useful to the public, but it can also place additional demands on the formal response effort for additional verification.

Citizen communications can also create new opportunities for the creation of new, temporary organizations that help with the informal response effort. The idea of emergent or ephemeral organizations that arise following disaster is not at all new; in fact, it is one of the hallmarks of disaster sociology, and supports the need for communities to be able to improvise responses under uncertain and dynamic conditions (Lanzara 1983; Mendonça and Wallace 2004; Wachtendorf 2004; Wachtendorf and Kendra 2005). ICT-supported communications, however, add another powerful means by which this kind of organization can occur. No longer do people need the benefit of physical proximity to coordinate and discover each other serendipitously.

These forms of public participation map to functions of the formal response effort: the strategic or intelligence functions, public relations, and the coordination of relief work. How to account for the role of public participation in formal response efforts, however, is not an easy question.

Li et al. (2012) suggested that ICT be used to manage national, regional, natural, and disasters caused by humans. It is useful during all the different stages of an emergency, including emergency prevention, mitigation, preparedness, emergency response, and emergency recovery. ICT technologies can be used for (a) effective warning of emergencies using different communication channels; (b) integrating information about necessary supplies and other sources; (c) coordinating disaster relief work; (d) encouraging social, institutional, and public responses; and (e) evaluating the damages caused by a disaster and the need for disaster relief.

3.9.12 RFID (Radio Frequency Identification)

RFID (Radio Frequency Identification) is an automated identification technology that uses radio frequency waves to transfer data between a reader and items that have RFID devices affixed to them. "Smart labels" contain a microchip and antenna and operate at internationally recognized standard frequencies. Similar to a bar code, the RFID tag stores data but offers enhanced data collection and significant advantages such as being able to read without a direct view of the RFID label and a dynamic read/write capability.

A basic RFID system is a combination of three major components, namely a tag (active, passive, or semi-passive) that serves as an electronic database and can be attached to or embedded in a physical object; a reader and its antennae that communicate with the tag without requiring a line of sight; and a host server equipped with middleware to manage the RFID system and interact with intra- and inter-organizational information systems.

RFID tags are classified by size and functionalities such as power source, operating frequency, data storage capacity and capability, operational life, and cost. For example, active tags have a tiny embedded battery as a power source and passive tags contain no power source. Semi-passive tags function in the same way as a passive tag; however, they have a power source that allows them to use an on-board sensor to monitor their environment; therefore, they are better suited to cold supply chain monitoring. Low-frequency tags use frequencies ranging from 125 to 134 kHz, high-frequency (HF) tags use the 13.56 MHz frequency, and ultra-high-frequency (UHF) tags use frequencies from 866 to 960 MHz. Microwave tags operate with frequencies ranging from 2.4 to 5.8 GHz. Data storage capacity and capability vary; RFID tags may be read-only or read/write. The data transmission rates of active tags are higher than passive tags, and the data-storage capacity of passive tags is smaller than active tags. Active tags' operational life is usually shorter than that of passive tags. Finally, passive tags are less expensive than active or semi-active tags as they do not require a power source. This fact positions passive tags as the best solution for RFID-enabled supply chain applications with decreasing costs (Véronneau and Roy 2009).

In addition, RFID tags are made in a variety of sizes and can be applied to or embedded into many products. In general, RFID active tags are much bigger than passive tags.

RFID enables mobile devices and tags are used for posting, gathering, storing, and sharing building assessment information efficiently with few errors, leading to improved efficiency and effectiveness in the emergency response process.

3.9.13 Internet of Things

The Internet of Things is a self-configured dynamic global network infrastructure with standards and interoperable communication protocols where physical and virtual "things" have identities, physical attributes, and virtual personalities, and are integrated seamlessly into information infrastructure (European Commission 2009). The concept of "things" in the network infrastructure refers to any real or virtual participating actors such as real world objects, human beings, virtual data, and intelligent software agents.

The purpose of the Internet of Things is to create an environment in which basic information from any one of the networked autonomous actors is shared efficiently with others in real time. Such a vision is possible with more powerful and efficient data collection and sharing ability. The Internet of Things is capable of supporting sophisticated decision support systems by providing services in a more accurate, detailed, and intelligent manner. While workflow task descriptions, task constraints (including needed resources), and relationships between tasks is static in many applications, the constantly changing environment and requirements during an emergency require an ability to alter the workflow rapidly and accurately (Wang et al. 2008). The Internet of Things, with its potential for instantaneous updates of status, requirements, and other information, enables dynamic workflow adaptations.

The fundamental characteristics of the Internet of Things technology are: (1) the Internet of Things is a global and real-time solution; (2) it is primarily wireless oriented and able to provide comprehensive data about its surroundings in both indoor and outdoor environments; and (3) it has the ability to monitor the environment remotely and trace or track objects.

The first fundamental characteristic is that it is a global and real-time solution. First, because the Internet of Things technology is Internet- or other wide-area network-based, the scope of the Internet of Things has no physical boundary. Any object linked with the network can be incorporated into the Internet of Things.

Second, data communication is real time or almost real time over the Internet of Things, which is different from traditional databases or web systems.

The second characteristic is that it is wireless and can provide comprehensive data about its surroundings. RFID sensor networks in the Internet of Things integrate RFID networks and wireless sensor networks into a unified information infrastructure. No line of sight is required for sensing in RFID sensor networks. This feature significantly increases the richness of information.

The third characteristic is its ability to monitor the environment and trace and track objects. By combining the use of RFID sensor networks with other technologies such as GPS or infrared sensor detection, RFID sensor networks have the capacity for wireless, real time monitoring and tracking of any tagged object in an indoor or outdoor environment to provide complete visibility of resources. Such visibility enables an instant response to any exception event, distributed information sharing among multiple organizations and multiple users, and resource distribution.

The Internet of Things technology can enhance emergency response from the following four perspectives.

(1) Accountability of resources and personnel. Accurate and real-time accountability ensures that there is an accurate count of resources and personnel on the scene or on the way to the scene. A lack of accountability can lead to dangerous situations when it is not recognized that resources or personnel are missing (Jiang et al. 2004). The Internet of Things technology provides a global and real time solution for monitoring, tracing, and tracking resources and personnel on the scene, even inside a premise or on the way to the scene. This solution works in indoor environments with poor visibility as well as outdoor environments with high visibility (Ramirez et al. 2009). Therefore, the Internet of Things technology offers emergency response accurate and real time accountability of resources and personnel.

(2) Situation assessment. Enhanced situational awareness leads to better decision-making in emergency response operations. The Internet of Things technology possesses the ability to provide real-time and comprehensive data about the incident scene via wireless sensor networks, RFID, and other techniques. Therefore, fast and accurate situational awareness can be achieved by gathering these comprehensive data and presenting them to emergency personnel (Wickler and Potter 2009).

(3) Resource allocation. Effective emergency response operations rely on sufficient supplies of responding personnel and resources (Yang et al. 2011). The Internet of Things technology provides for visibility of response personnel and their remaining resources through its ability to monitor, trace, and track remotely. Therefore, resources can be allocated most efficiently and delivered to the disaster scene. Furthermore, resource allocation increases the capability of emergency response operations by making limited resources available to more emergency response operations. This finding extends the Internet of Things application from primarily the management of the logistics supply chain to dynamic resource allocation.

(4) Multi-organizational coordination. Emergency response operations require the participation of a wide range of organizations, including fire brigades, police forces, ambulance services, local or national public sectors, and humanitarian aid organizations, etc. Extensive information and resource sharing between separate organizations is crucial to the success of emergency response operations (Yang 2007). The Internet of Things technology provides real-time information on disaster development and the remaining resources of each participating organization, as well as an information-sharing infrastructure. Multi-organizational coordination is well supported by the rich information provided by the information infrastructure of the Internet of Things.

Emergency response operations enhanced by the above four perspectives can achieve three aspects of strategic value: efficient cooperation among various organizations, accurate situational awareness, and complete visibility of response forces and their remaining capabilities. Strategic value is realized in three ways: information sharing, information retrieval, and information explanation, contributed to by the characteristics of the Internet of Things technology.

References

Akgün V, Erkut E, Batta R (2000) On finding dissimilar paths. Eur J Oper Res 121:232–246

Alhajraf S, Al-Awadhi L, Al-Fadala S, Al-Khubaizi A, Khan AR, Baby S (2005) Real-time response system for the prediction of the atmospheric transport of hazardous materials. J Loss Prev Process Ind 18(4–6):520–525

Al-Natour S, Benbasat I, Cenfetelli R (2011) The adoption of online shopping assistants: perceived similarity as an antecedent to evaluative beliefs. J Assoc Inf Syst 12(5):347–374

Amailef K, Lu J (2011) A mobile-based emergency response system for intelligent m-government services. J Enterp Inf Manag 24(4):338–359

Boyd D, Ellison NB (2007) Social network sites: definition, history, and scholarship. J Comput Mediated Commun 13(1)

Cacas M (2010) Feds expand virtual worlds use. FederalNewsRadio.com. Retrieved 30 Oct 2011, from http://www.federalnewsradio.com/?nid=697&sid=1957088

Caliper (2011) TransCAD overview. Available online at http://www.caliper.com/tcovu.htm. Last accessed on 15 Feb 2011

Campbell D (2002) 9/11: a healthcare provider's response. Front Health Serv Manag 19(1):3–13

Chang N-B, Wei YL, Tseng CC, Kao C-YJ (1997) The design of a GIS-based decision support system for chemical emergency preparedness and response in an urban environment. Comput Environ Urban Syst 21(1):67–94

Cheng S, Chen G, Chen Q, Xiao X (2009) Research on 3D dynamic visualization simulation system of toxic gas diffusion based on virtual reality technology. Process Saf Environ Prot 8(7):175–183

Daley E (2003) Wireless interoperability. Publ Manag 85(4):6–10

Dantas A, Ferreira F (2010) Prioritisation and deployment of physical and human resources during disasters. In: Proceedings of T-LOG conference 2010, 6–8 Sept, Fukuoka, Japan

DeLaurentis D, Han E, Kotegawa T (2008) Network-theoretic approach for analyzing connectivity in air transportation networks. J Aircr 45(5)

Deneubourg J-L, Aron S, Goss S, Pasteels J-M (1990) The self-organizing exploratory pattern of the Argentine ant. J Insect Behav 3(1990):159

Dijkstra EW (1959) A note on two problems in connection with graphs. Numer Math 1:269–271

Dijkstra E, Misa TJ (ed) (2010-08) An interview with Edsger W. Dijkstra. Commun ACM 53(8):41–47

Eppstein D (2011) Finding the K-shortest paths, 1997. Available online at http://www.ics.uci.edu/~eppstein/pubs/Epp-SJC-98.pdf. Last accessed on 15 Feb 2011

European Commission (2009) Internet of things strategic research road map electronic text at http://www.internet-of-things-research.eu/pdf/IoT_Cluster_Strategic_Research_Agenda_2009.pdf. Last accessed in Feb 2012

Ferreira F (2010) Dynamic response recovery tool for emergency response within state highway organizations in New Zealand. Doctor of philosophy thesis, University of Canterbury, Christchurch, New Zealand

Fiedrich F, Burghardt P (2007) Agent-based systems for disaster management. Commun ACM 50 (3):41–42

Fry EA, Lenert LA (2005) MASCAL: RFID tracking of patients, staff and equipment to enhance hospital response to mass casualty events. In: AMIA annual symposium proceedings, vol 2005, p 261. American Medical Informatics Association

Goss S, Aron S, Deneubourg J-L, Pasteels J-M (1989) Self-organized shortcuts in the Argentine ant. Naturwissenschaften 76(1989):579–581

Griffin L, de Leaster E (2009) Social networking healthcare. In: IEEE 6th international workshop on wearable micro and nano technologies for personalized health (pHealth), pp 75–78

Hamagami T, Hirata H (2003) Method of crowd simulation by using multi-agent on cellular automata. In: Proceedings of IEEE/WIC international conference on intelligent agent technology (IAT'03)

Henley EJ, Kumamoto H (1992) Probabilistic risk assessment: reliability engineering, design and analysis. Institute of Electrical and Electronics Engineers, New York, USA

Huang C-C, Tseng T-L (2004) Rough set approach to case-based reasoning application. Expert Syst Appl 26(3):369–385

Jain S, McLean CR (2005) Integrated gaming and simulation architecture for incident management training. In: Kuhl ME, Steiger NM, Armstrong FB, Joines JA (eds) Proceedings of the 2005 winter simulation conference. Institute of Electrical and Electronics Engineers, Piscataway, New Jersey, pp 904–913

Jiang X, Hong JI, Takayama LA, Landay JA (2004) Ubiquitous computing for firefighters: field studies and prototypes of large displays for incident command. In: Proceedings of the SIGCHI conference on human factors in computing systems, Vienna, Austria, 2004, pp 679–686

Kwan M-P, Lee J (2005) Emergency response after 9/11: the potential of real-time 3D GIS for quick emergency response in micro-spatial environments. Comput Environ Urban Syst 29(2):93–113

Lai C (2011) Network terminal reliability. Master of engineering in transportation research report. Department of Civil and Natural Resources Engineering, University of Canterbury, Christchurch, New Zealand

Lanzara GF (1983) Ephemeral organizations in extreme environments: emergence, strategy, extinction. J Manag Stud 20(1):71–95

Lei XE, Zhang MG, Han ZW (1998) Numerical prediction basis and models of air pollution. Weather Press, Beijing, pp 223–240

Li Z, Ramani K (2007) Ontology-based design information extraction and retrieval. Artif Intell Eng Des Anal Manuf 21(2):137–154

Li J, Li Q, Liu C, Khan SU, Ghani N (2012) Community-based collaborative information system for emergency management. Comput Oper Res, 1–9

Liao Z, Wang B, Xia X, Hannam PM (2012) Environmental emergency decision support system based on artificial neural network. Saf Sci 50(1):150–163

Lin C (2000) Random petri net and system performance analysis. Tsinghua University Press, Beijing

Liu X, Li W, Tu YL, Zhang WJ (2011) An expert system for an emergency response management in networked safe service systems. Expert Syst Appl 38(9):11928–11938

Macal CM, North MJ (2010) Tutorial on agent-based modeling and simulation. J Simul 4 (3):151e62

McGrath D, Hunt A, Bates M (2005) A simple distributed simulation architecture for emergency exercises. In: Ninth IEEE international symposium on distributed simulation and real time applications

McKnight DH, Choudhury V, Kacmar C (2002) Developing and validating trust measures for e-commerce: an integrative typology. Inf Syst Res 13(3):334–359

Mendonça D, Wallace WA (2004) Studying organizationally-situated improvisation in response to extreme events. Int J Mass Emergencies Disasters 22(2):5–29

National Research Council (2002) Making the nation safer: the role of science and technology in countering terrorism. National Academies Press, Washington DC, USA

Nicholson A (2007) Optimising network terminal reliability. In: Van Zuylen HJ (ed) Proceedings of the 3rd international symposium on transport network reliability, The Hague, Netherlands, 19–20 July 2007

Nicholson A, Giovinazzi S (2010) Hazard loss estimation and transport network reliability. In: Proceedings of the 4th international symposium on transport network reliability, 22–23 July 2010, University of Minnesota, Minneapolis, USA

Qing Y, Huimin MA, Yanling YU (2012) Multi-agent risk identifier model of emergency management system engineering based on immunology. Syst Eng Procedia 4:385–392

Ramirez L, Denef S, Dyrks T (2009) Towards human-centred support for indoor navigation. CHI, Boston, USA, pp 1279–1282

Rongxi C (1997) The capacity evaluation of shipboard command and control system based on Petri net. Syst Eng Electron Technol 5:18–24

Sang JG, Wen SG (1992) Numerical calculation of atmospheric diffusion. Weather Press, Beijing, pp 140–149

Scott DW (2011) Using social media in engineering support and space flight operations control. In: IEEE automatic control, pp 1–14

Scott DM, Novak DC, Aultman-Hall L, Guo F (2006) Network robustness index: a new method for identifying critical links and evaluating the performance of transportation networks. J Transp Geogr 14(3):215–227

Sharafat AR, Ma'rouzi OR (2009) All-terminal network reliability using recursive truncation algorithm. IEEE Trans Reliab 58(2)

Shendarkar A, Vasudevan K, Lee S, Son Y-J (2008) Crowd simulation for emergency response using BDI agents based on immersive virtual reality. Simul Model Pract Theory 16(2008):1415–1429

Sheu J-B (2007) Challenges of emergency logistics management. Transp Res Part E 43(2007):655–659

Taniguchi E, Ferreira F, Nicholson A (2012) A conceptual road network emergency model to aid emergency preparedness and response decision-making in the context of humanitarian logistics. Procedia Soc Behav Sci 39(2012):307–320

Taylor DG, Lewin JE, Strutton D (2011) Friends, fans, and followers: do ads work on social networks? How gender and age shape receptivity. J Advertising Res 51(1):258–275

Tran Quoc D, Kameyama W (2007) A proposal of ontology-based health care information extraction system: VnHIES. In: IEEE international conference on research, innovation and vision for the future, pp 1–7

Véronneau S, Roy J (2009) RFID benefits, costs, and possibilities: the economical analysis of RFID deployment in a cruise corporation global service supply chain. Int J Prod Econ 122(2):692–702

Victor AN, Boris PF (1996) Consideration of node failures in network-reliability calculation. IEEE Trans Reliab 45(1):127–128

Wachtendorf T (2004) Improvising 9/11: organizational improvisation following the world trade center disaster. Dissertation, University of Delaware Disaster Research Center

Wachtendorf T, Kendra JM (2005) Improvising disaster in the city of Jazz: organizational response to Hurricane Katrina. Understanding Katrina: Perspectives from the Social Sciences. understandingkatrina.ssrc.org

Wakabayashi H (2004) Network reliability improvement: probability importance and criticality importance. In: Proceedings of second international symposium on transport network reliability

Wang J, Rosca D, Tepfenhart W, Milewski A, Stoute M (2008) Dynamic workflow modeling and analysis in incident command systems. IEEE Trans Syst Man Cybern Part A Syst Hum 38(5):1041–1055

Wang Y, Luangkesorn KL, Shuman L (2012) Modeling emergency medical response to a mass casualty incident using agent based simulation. Socio-Econ Plann Sci, 1–10

Wenger E (1998) Communities of practice: learning, meaning and identity. Cambridge University Press, Cambridge

Wickler G, Potter S (2009) Information-gathering: from sensor data to decision support in three simple steps. Inf Syst J 3(1):1–42

Yang L (2007) On-site information sharing for emergency response management. J Emerg Manag 5(5):55–64

Yang H-L, Wang C-S (2008) Two stages of case-based reasoning—integrating genetic algorithm with data mining mechanism. Expert Syst Appl 35(1–2):262–272

Yang L, Prasanna R, King M (2009) On-site information systems design for emergency first responders. J Inf Technol Theory Appl 10(1):5–27

Yang HJ, Yang L, Yang SH (2011) Hybrid Zigbee RFID sensor network for humanitarian logistics centre management. J Netw Comput Appl 34(3):938–948

Yates D, Paquette S (2010) Emergency knowledge management and social media technologies: a case study of the 2010 Haitian earthquake. Int J Inf Manag 31(1):6–13

Yuan Y, Detlor B (2005) Intelligent mobile crisis response systems. Commun ACM 48(2):95–97

Zeeshan A, Peña-Mora F, Chen A (2009) Supporting urban emergency response and recovery using RFID-based building assessment. Disaster Prev Manag 18(1):35–48

Zhong M, Shi C, Fu T, Hea L, Shi J (2010) Study in performance analysis of China urban emergency response system based on petri net. Saf Sci 48:755–762

Chapter 4
Case Study

This chapter provides two case studies about emergency response decision support that illustrate typical cases. The detailed case analysis creates a realistic model of the emergency response and demonstrates solutions to problems that have great value in practical applications.

4.1 Disaster Rescue Decision Support System for Beijing

Beijing is the political, economic, and cultural center of China and a domestic and international exchange center. The resident population in Beijing was over 20 million by the end of 2015. Disasters such as SARS, floods, droughts, blazes, and traffic accidents occur frequently in Beijing. Rescue, social security, command, and coordination work is particularly important in response to these sudden disasters. We conducted research into and developed a disaster rescue decision support system in Beijing in order to construct relevant contingency plans promptly, undertake effective coordination, command, and emergency rescue work, improve emergency response capacity to deal with unexpected disasters, and to reduce the adverse effects and loss of life and property from them.

We made use of information, communications, simulation, system optimization, and emergency response decision support technologies, with the emergency disaster rescue and security work in Beijing as the object. We took into account the Twelfth Five-Year Plan of Beijing, the General Plan of Beijing City (2004–2020), the Emergency Rescue Plan of Urban and Rural Residents Basic Consumer Goods Price Fluctuation in Beijing, the Victims Relief Work Contingency Plan of Beijing (for trial implementation), the Contingency Plan to Deal with the Public Emergency of the Beijing Civil Affairs Bureau, and other laws and regulations. The system we

© The Author(s) 2017
S. Shan and Q. Yan, *Emergency Response Decision Support System*,
SpringerBriefs in Business, DOI 10.1007/978-981-10-3542-5_4

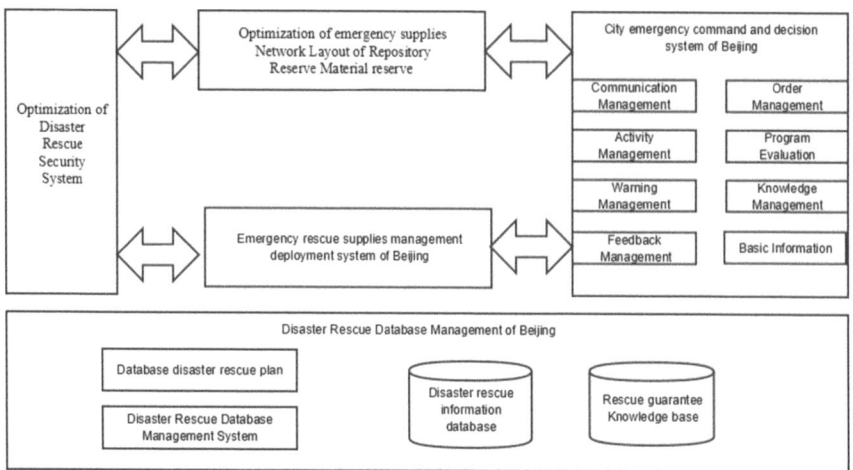

Fig. 4.1 Overall framework of the Beijing disaster rescue decision support system

devised optimizes the disaster relief and security system of Beijing and establishes a decision support system for disaster rescue in Beijing. The system assists in allocating and storing rescue resources before a disaster, providing effective support to disaster coordination and command, scheduling and distributing rescue resources after a disaster, and improving the system's ability to deal with unexpected disasters and rescue work in Beijing. The overall framework of the Beijing disaster rescue decision support system is shown in Fig. 4.1.

The Beijing disaster rescue decision support system has five components: optimizing Beijing's disaster rescue support system, optimizing the network layout and material reserves of emergency material storage, establishing Beijing's emergency rescue supplies management distribution system, establishing the city's emergency command and decision system, and establishing Beijing's disaster rescue database.

Optimizing Beijing's disaster rescue security system is based on disaster emergency management theory and methods, information technologies and means, a management optimization model, and strategy theory and technology. We analyzed and diagnosed the current disaster rescue and security system of Beijing in various ways from many different perspectives. Our proposed solution undertakes simulation analysis of the critical emergency plan, solves problems, improves relevant contingency plans, plans the disaster rescue and security technology system, and provides the methods and program support for the establishment of a unified disaster rescue decision support system for Beijing. Based on our analysis of the needs of the disaster rescue security system and the status of the current system, there are five areas needing optimization analysis to solve outstanding and urgent problems in the disaster rescue security system. These problems include improving the current disaster rescue and security technology system; simulation and optimization of key plans; planning the disaster rescue and security technology system;

and enhancing unified early warning, command, coordination, and management. Other needed work includes planning the disaster rescue materials management system; establishing the location, storage, and distribution of rescue supplies; scheduling optimization models and regulations; planning the disaster rescue and security knowledge system; enhancing the development and utilization of knowledge in disaster rescue; planning the performance evaluation system for disaster rescue and security; establishing the optimization model; and improving the efficiency of disaster rescue and security work.

Beijing's emergency rescue supplies management and deployment system is an important part of the city's disaster rescue decision support system. We analyzed the key features and all aspects of the current situation, including storage, procurement, deployment, and information processing of emergency supplies after disasters, to identify the main factors that affect the efficient deployment of emergency rescue supplies. We extracted and analyzed in detail the functional requirements of the emergency rescue supplies management and deployment system, and designed Beijing's emergency rescue supplies management and deployment system to ensure it has the capacity to supply emergency rescue materials accurately and efficiently. The emergency rescue supplies management and deployment system employs scientific management of reserve rescue supplies and decision support for emergency supplies deployment after disasters. It improves the timeliness supplies information for decision-makers and has the capability to analyze relevant data and estimate the role a variety of materials play in rescue results.

The dynamic acquisition of information, and statistical and information analysis about the number, types, and distribution of emergency rescue supplies etc., are important functions of the Beijing disaster rescue decision support system. It is the basis for supply management and deployment, plays a part in collecting basic information and the deployment and decision support for emergency supplies, maximizes the use of limited resources, and enhances the capabilities of disaster emergency supplies storage and distribution. The system uses RFID or other electronic tag information storage and marking functions with GPS satellite navigation and positioning technology, for monitoring and command of emergency rescue supplies storage and freight vehicles to form a combination "point-line-plane" and real time weather information network. Seamless integration of GIS and GPS location information allows the system to position freight vehicles precisely, track dynamically, control processing, and provide visualization management. GPRS enables communication across the entire regulatory platform to achieve real time monitoring and management of disaster rescue and disaster mitigation, achieve timely and reliable alarm and emergency calls for freight route deviation, freight vehicles pause timeout, and other abnormal conditions. It ensures the efficacy and safety of rescue supply delivery.

The city emergency command management system is the system engineering core component of the whole emergency response process, covering all emergency monitoring and control, forecasting and early warning, alarm, disposal, rehabilitation, reconstruction, and other aspects, including the business and technology systems which are two fundamental aspects of emergency management.

"Emergency management system" usually denotes a business case covering three systems, including contingency plans, the emergency response system, operations, and the rule of law. Technology systems for emergency management command comprise subsystems for basic disaster information, early warning management, emergency response communication, emergency response command management, decision-making program evaluation, emergency response activity management, emergency decision feedback, knowledge management, and so on.

Disaster rescue database management in Beijing includes overall planning, the design of Beijing Emergency Aid database management, and the building disaster rescue and knowledge base, which supports smooth disaster rescue and security in Beijing by providing the necessary data and knowledge for rescue work. The disaster rescue database of Beijing includes five categories: database content (the basic information database), and geographic information, disaster rescue information, disaster rescue experts, and disaster rescue security knowledge databases. The basic information database includes the meteorological, population distribution, roads and bridges, address, property, safe distance, land, water conservancy facilities, and devices and equipment databases. Geographic Information Database includes databases for geographic information spatial data interface and road space; spatial databases for dangerous sources, protection objectives, and disaster sites; and a hedge position database. The disaster rescue information database includes databases for firefighting forces, distributed defense, medical information, and evacuation.

4.2   Beijing Flood Emergency Management Intelligent Database System

Flood is one of the most frequently occurring and the most serious natural disasters in the history of Beijing. On July 21–22, 2012, devastating floods in Beijing affected 1.602 million people. Economic losses totaled 11.64 billion and 10,660 houses collapsed. The death toll was at least 79 in Beijing and surrounding areas. There are puzzles and challenges in flood rescue: how to develop a scientific and rational flood rescue plan quickly? How to change the textbook flood case into a structured knowledge-based case? How to use the experience of the previous city flood rescue plan automatically? How to identify flood rescue knowledge that has not been used before in many urban floods? How to improve the city's flood emergency rescue response capability? Beijing's flood emergency management intelligent database system conducted significant exploratory work in these areas. The system's concepts and framework are shown in Fig. 4.2.

The first step in constructing ontologies is to determine the entity's scope in order to determine the concepts, attributes, and relationships in the entity's field. The strategy and tactics of the flood emergency database domain ontology is based on types of floods; floods' locations, disaster levels, affected areas, and intensity

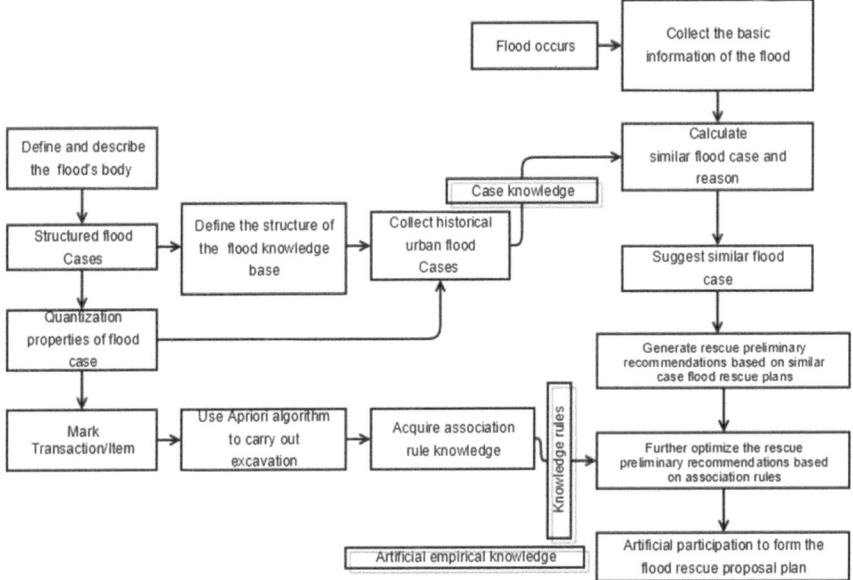

Fig. 4.2 Framework of Beijing's flood emergency management intelligent database system

(number of old houses collapsed, number of people affected, mountain landslides, rainfall, and reservoir water level); and the causes and duration of previous disasters. The strategic and tactical dimensions of flood emergency response include rescue teams (transport sector, publicity departments, emergency command center, rescue department, other rescue teams) and rescue resources (emergency equipment, relief supplies, and materials for daily living). Strategy and tactics for flood emergency restoration and reconstruction includes insurance and compensation, disaster rescue, materials supplementary, facilities repair and so on.

From the perspective of ontological classification analysis, the flood emergency response entity should include top-level, domain, task, and application ontologies simultaneously. The domain ontology should include geography, meteorology, rescue teams, materials and so on.

We screened the content analysis of disaster emergency rescue, lessons from earlier studies, and terminologies in the field, in accordance with the classification system of concepts in the field and the composite structures of concepts, to obtain five flood emergency database fields: flood type, flood property, flood phenomenon, disaster rescue, recovery, and reconstruction.

Flood type is qualitative—flood belongs to one of the categories of city flooding disasters, floods in mountain areas, reservoir and river floods, or suburban or town flooding disasters. Flood property concerns basic flood information, comprising affected areas, place, time, and disaster rating. Flood phenomenon concerns particular information about individual floods, comprising disaster strength (damaged houses, number of people affected, water depth on the road, mountain landslide, rainfall, and reservoir water level) and the causes of the disaster and its duration. Disaster rescue is the emergency response after the floods, including rescue resources (emergency equipment, relief supplies, and materials for daily living) and rescue teams (transportation department, publicity department, meteorological department, communications department, etc.).

Our analysis of flood rescue areas identifies that these concepts are hierarchical, structured, and formal, allowing us to construct an ontology model of the flood emergency database domain. In line with the concept of a hierarchical system, we defined the five categories listed above. We provide concrete examples of each category in Table 4.1.

The Beijing flood emergency management intelligent database system uses knowledge reasoning of the Beijing flood emergency database based on Apriori algorithms. An Apriori algorithm is a frequent item sets algorithm that mines association rules. It is often used to mine frequent item sets by two stages of candidate set generation and closed down plot detection. The algorithm's basic approach is first, to find all frequent sets. The appearance frequency of these item sets is at least is the same within the predefined minimum support degree. Next,

Table 4.1 Beijing flood emergency database domain ontology classes

First tier category	Second subclass	Third subclass	Specific meaning
Flood type	Flooding disasters in cities		Includes overflowing river, wide water area, traffic gridlock, collapse of damaged houses, water in underground facilities, lack of power and water, and other secondary disasters
	Mountain disasters		Includes flash floods, mudslides, landslides, collapse of a large area and so on
	Reservoir and river floods		Includes cracked, collapsed, or overflowing piping; burst or collapsed dike, submerged gates, or river water level rise
	Suburban and town flooding disasters		Suburban counties including cities and towns (including Metro) and flooding caused by secondary disasters
	Affected areas		Areas directly or indirectly affected by flooding
	Location		Principal place of flood occurrence

(continued)

Table 4.1 (continued)

First tier category	Second subclass	Third subclass	Specific meaning
Flood property	Time		Time of flood occurrence
	Disaster levels		From low to high, divided into general flood prevention emergency (level IV), larger flood control emergency (level III), major flood control emergency (level II), and major flood control emergency (level I)
	Disaster strength	Precipitation Number of people affected Economic losses	Specific flood precipitation, number of people affected, and economic losses
	Causes of the disaster		Internal or external flood factors
Flood phenomenon	Duration of the disaster		Length of time of flood impact.
	Rescue resources	Materials for daily living, rescue supplies, emergency equipment, etc.	Materials for daily living, rescue supplies, emergency equipment, and other supplies in flood rescue
	Rescue teams	Directing department, propaganda department, transportation department, etc.	Directing department, publicity department, transportation department, and other rescue teams
Disaster rescue	Insurance and compensation		After the floods, flood damage facilities, equipment, and loss of life and property in the affected areas
	Disaster rescue		Timely deployment of relief funds and materials, organizing resettlement of affected people, improving disaster victims' temporary living arrangements, organizing affected people, and reconstructing collapsed houses

(continued)

Table 4.1 (continued)

First tier category	Second subclass	Third subclass	Specific meaning
Recovery and reconstruction	Supplementary material		Current flood control materials consumption, according to the classification of the financing of supplementary requirements
	Facilities repair		Flood water conservancy, municipal facilities, transportation, electricity, telecommunications, oil, gas, water supply, drainage, housing, civil air defense projects, river-crossing pipelines, hydro facilities, tissue repair and restore function

strong association rules generate frequent sets, which must meet minimum support degrees and minimum confidence intervals. The rules produced by the frequent sets in the first step are used to produce all the rules that contain only items where the right side of each rule has a single item. This is the definition of the rule. Once these rules are generated, only frequently used minimum credibility rules remain. We then employ the recursive method to generate all frequency sets. In the example, we represent structured knowledge of Beijing flood events based on the ontology knowledge model, and every ontology model attribute is represented as an initialization item in an Apriori algorithm, namely item sets. We value discrete items in the Apriori algorithm directly, but determine the discrete value of continuous items based their purpose. For example, the flooding date is a continuous variable. Floods occur more frequently in the rainy season, so the flooding date is into "rainy season" and "non-rainy season." According to information from the Beijing Civil Affairs Bureau, Beijing's rainy season is June, July, and August. Therefore, we define floods that occur in June, July or August as "rainy season," while floods that occur in other months are defined as "non-rainy season."

The Beijing flood emergency management intelligent database system integrates Euclidean distance and cosine distance to build a similarity computation model of flood event and case studies. We measured the similarity of emergencies and case studies across the two dimensions of absolute numbers and relative proportions, to produce a more scientific and rational analysis of cases similar to the current flood event, providing strong support for the development of scientific decision-making. Similarity is based on the distance between two points in Euclidean space.

Assuming that X, Y are two points of n-dimensional space, the Euclidean distance between them is:

$$D(X,\ Y) = \sqrt[2]{\sum (x_i - y_i)^2} \tag{4.1}$$

As can be seen, when n = 2 the Euclidean distance is the distance between two points on the same plane. The following formula is used widely for conversion when identifying similarity: the smaller the distance, the greater the similarity.

$$sim(X,\ Y) = \frac{1}{1 + d(X,\ Y)} \tag{4.2}$$

Principle: the similarity s is defined by Euclidean distance d, s = 1/(1 + d).

Range: [0, 1]. The greater the value, the smaller the d. That is to say, the closer the distance, the greater the similarity.

Cosine similarity is widely used to calculate the similarity of document data:

$$T(X,\ Y) = \frac{x \cdot y}{\|x\|^2 \times \|y\|^2} = \frac{\sum x_i y_i}{\sqrt[2]{\sum x_i^2}\sqrt[2]{\sum y_i^2}} \tag{4.3}$$

Principle: cosine value is calculated as the angle formed between two points in multi-dimensional space and the set point.

Range: [−1, 1]. The larger the value, the larger the angle. The farther apart the two points are, the less the similarity.

In short, by using current knowledge representation, knowledge mining, and other key research technologies, the Beijing flood emergency management intelligent database system combines ontology, reasoning, and association rules to construct flood intelligent database systems. It employs a human-computer interaction subsystem of decision support, joining human experience to the rescue plan to assist decision support. The system contains three main components: reasoning, association rules, and human-computer interaction. These three parts interact to form a framework for flood cases, including flood descriptions and a two-part rescue program. Case histories are collected and structured case descriptions created to obtain case features. Feature attributes are quantified and the text description converted to properties of rule mining. Characteristic flood and rescue properties are input as Apriori algorithms, and association rules between the two properties are extracted. When floods occur, flood-related information is collected and flooding attributes extracted according to the defined organizational structure in the case library. The similarity algorithm is used to identify floods that most closely resemble the current flood and the current rescue plan is reused in the case library to create preliminary rescue recommendations. Rules of association are applied to the current flood description to optimize the initial rescue plan. Finally, the

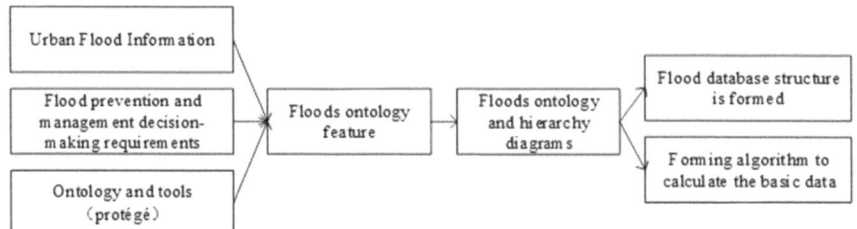

Fig. 4.3 Flood emergency management flowchart

recommended rescue plan is shown to users, and the human-computer interaction system provided by the decision support system allows users to modify the cases, after which the cases as adjusted by the users will be deposited as the final case into the Case Base (Fig. 4.3).